OECD Studies on Water

Enhancing Water Use Efficiency in Korea

POLICY ISSUES AND RECOMMENDATIONS

OECD
BETTER POLICIES FOR BETTER LIVES

This work is published under the responsibility of the Secretary-General of the OECD. The opinions expressed and arguments employed herein do not necessarily reflect the official views of OECD member countries.

This document, as well as any data and any map included herein, are without prejudice to the status of or sovereignty over any territory, to the delimitation of international frontiers and boundaries and to the name of any territory, city or area.

Please cite this publication as:
OECD (2017), *Enhancing Water Use Efficiency in Korea: Policy Issues and Recommendations*, OECD Studies on Water, OECD Publishing, Paris.
http://dx.doi.org/10.1787/9789264281707-en

ISBN 978-1-78040-938-2 (print)
ISBN 978-92-64-28170-7 (PDF)

Series: OECD Studies on Water
ISSN 2224-5073 (print)
ISSN 2224-5081 (online)

The statistical data for Israel are supplied by and under the responsibility of the relevant Israeli authorities. The use of such data by the OECD is without prejudice to the status of the Golan Heights, East Jerusalem and Israeli settlements in the West Bank under the terms of international law.

Photo credits: Cover © K-water, "Soyanggang Multipurpose Dam, K-water"

Corrigenda to OECD publications may be found on line at: *www.oecd.org/about/publishing/corrigenda.htm*.

Note from the Secretariat

This report was prepared by the OECD Secretariat to support discussions towards sustainable management of water scarcity in Korea. It lays down assessments and recommendations, building on interviews conducted with a range of Korean stakeholders during the fact-finding mission (4-8 April 2016), information provided by the Korean authorities (MoLIT and K-water) and technical discussions (19-21 October 2016). Draft assessments and recommendations were discussed with several ministries and a range of stakeholders during a policy seminar organised in Daejeon, on 2-3 March 2017. Comments received during and after the seminar have been reflected in the current document.

The report includes inputs from an expert team of the Toulouse School of Economics (Henrik Andersson, Hyelin Jeon, Céline Nauges and Arnaud Reynaud). Amir Shakarov (Director Economics Division, Governmental Authority for Water and Sewage from Israel), Leonid Maryakhin (Manager of civil and environment engineering, Department of sewage systems development, Israel Water Authority) and Jan Leentvaar (former Director Chief-Inspector of the Water Management Inspectorate of the Netherlands) contributed throughout the project as peer reviewers. They are gratefully acknowledged.

The report was edited by Caitlin Connelly. Janine Treves provided editorial suggestions and Peter Vogelpoel did the typesetting.

The recommendations in the report do not necessarily reflect the opinions of the stakeholders met during the policy dialogue. They are aspirational, recommended by the OECD, and not mandatory. The OECD hopes they can inspire new ways of enhancing water use efficiency in Korea, taking account of prevailing policies, the specificities of Korean practices and emerging challenges.

The OECD Secretariat remains committed to help address water-related issues in Korea and continue excellent co-operation with Korean authorities on these and related issues.

The project was fully funded through a financial contribution from MoLIT and K-water.

Table of contents

Figures

Tables

Boxes

Acronyms and abbreviations

ADB	Asian Development Bank
CEO	Chief Executive Officer
GDP	Gross Domestic Product
ICT	Information and Communication Technology
IWRM	Integrated Water Resources Management
K-SWM	Smart Water Management in Korea
KECO	Korea Environment Corporation
KEI	Korean Environment Institute
KHNP	Korea Hydro and Nuclear Power Corporation
KICT	Korea Institute of civil Engineering
KRC	Korea Rural Community Corporation
KRIHS	Korea Research Institute for Human Settlements
KRILA	Korea Research Institute for Local Administration
KRW	Korean Republic Won
KWI	Korea Waterworks Management Institute
MAFRA	Ministry of Agriculture, Food and Rural Affairs
ME	Ministry of Environment
MoI	Ministry of the Interior
MoLIT	Ministry of Land, Infrastructure and Transport
MoSF	Ministry of Strategy and Finance
MOTIE	Ministry of Trade, Industry and Energy
MPSS	Ministry of Public Safety and Security
NDMI	National Disaster Management Research Institute
NGO	Non-Governmental Organisation
NIER	National Institute of Environment Research
SWM	Smart Water Management
SWG	Smart Water Grid
WTP	Willingness To Pay

Executive summary

An opportunity to review the management of water quantity in Korea

Water management in Korea has been able to deal with and facilitate rapid urbanisation and economic growth. This remarkable achievement is a result of heavy investment in quantitative development strategies to meet water needs and protect against risks of floods and droughts.

Water management in Korea now faces three long-term trends, which challenge prevailing policy and investment responses: demographic trends, in particular rapid ageing of the population, will drive water demand and strain revenues; economic trends and fiscal consolidation will affect the capacity of public funds to finance water management; and climate change will have an impact on water availability and the exposure to water-related risks.

This context calls for a review of prevailing policy responses and an assessment of whether they are fit for future challenges. The review focuses on three sets of instruments which play a role in water quantity management in Korea: economic instruments that promote water efficiency use in Korea; Smart water Management and the combination of water and information and communication technology to manage water resources and to deliver water services; and water allocation regimes.

The process was co-ordinated by the Ministry of land, Infrastructure and Transport in Korea, with active support from K-water. The Ministry of Land, Infrastructure and Transport plays a central role in promoting water use efficiency in Korea. It is responsible for water quantity management and multi-regional water supply. It establishes policies for the sector, prepares plans for the sector's development, and oversees river basin management. The process involved a variety of stakeholders, from other administrations, civil society organisations and academia.

Strengths and limitations of three economic instruments in place

Three economic instruments contribute to water quantity management and potentially promote water use efficiency in Korea: i) a river water use fee is levied on water users (energy generators, industry, and domestic users) who abstract water from a river; ii) a dam water tariff is a uniform volumetric tariff levied on water users who abstract water from a river or a dam via a contract with K-water; and iii) multi-regional water tariffs are two-part tariffs set nationwide to recover the costs of supplying water through multi-regional systems; different tariffs apply to different water quality grades (raw water, settled water, purified water).

These instruments were not designed to promote water use efficiency or address scarcity. They do not generate sufficient revenue required to maintain and extend existing infrastructures. This represents a missed opportunity to manage water demand and to minimise future investment needs in water infrastructure to augment supply.

OECD analyses and international good practice suggest Korean authorities would benefit from adjusting these instruments so that they promote water use efficiency in Korea. First, tariffs should better reflect the cost of supply at the abstraction point. Second, revenues could finance expenditure programmes that effectively contribute to river maintenance in the basin. Moreover, in the longer term, an abstraction charge could be considered, that reflects the opportunity costs of using water in basins where water is scarce. Any step in this direction would benefit from a gradual, long-term, inclusive, staged approach that engage stakeholders at both national and local levels.

Options to expedite the deployment of Smart Water Management in Korea

Smart Water Management in Korea has the potential to contribute substantially to water use efficiency in Korea and abroad. It can support decisions about the management of dams and reservoirs, and allocation of water in the system. It can support value-adding services to water utilities (such as leak detection) and water users (such as real-time information on water quality), thereby enhancing the performance of water utilities in Korea and abroad.

However, the diffusion of Smart Water Management is hindered by several bottlenecks, in Korea and abroad. Nonetheless, they can be overcome by targeted measures, such as the promotion of an integrated approach to water management, tariffs that reflect the full cost of supplying water, or models that minimise needs for new data. Such measures should be accompanied by involvement of final water users in the definition of additional services that contribute to their needs and by capacity building in municipalities.

Water allocation regimes: A pending issue

Well-designed water allocation regimes can promote water use efficiency and discourage wastage. They can contribute to better water management by allocating water where it creates most value for the Korean community.

However, the capacity of water allocation regimes in Korea to deliver such outcomes and contribute to water use efficiency is hampered by the coexistence of water entitlements acquired before and after the construction of dams: this situation creates legal disputes and inequities as regards who can access water and under which conditions.

Reforming water allocation regimes can generate welfare gains in Korea. Now, these reforms are very challenging. International experience is useful to explore how such reforms can be managed, with appropriate accompanying measures. The ambition of the report in this regard is limited to flagging some of the limitations of the prevailing situations, and sharing some options to move this agenda forward. This is a conversation that is expected to go on in Korea.

Opportunities to strengthen water governance for water use efficiency

It is advised that the institutional framework be amended to promote water use efficiency in Korea. It should be able to reflect local conditions and engage with relevant stakeholders at national, basin, or local scale. Some recent developments along these lines should be encouraged: educate and inform the population and water users about water scarcity, the opportunity cost of misusing or misallocating water, and the cost of supplying water; develop a strategic programme towards stakeholder engagement, with result-oriented performance management; basin organisations have a role to play.

Assessment and recommendations

Korea, a water-stressed country, was able to grow and urbanise faster than most other countries, by investing in water infrastructure to augment supply and prevent flood risks. Long term trends such as an ageing population and more uncertainty about water availability – due to climate change – question the prevailing model, characterised by heavy reliance on (publicly financed) infrastructure.

This chapter summarises the main assessments and recommendations that come out of the policy dialogue with Korean stakeholders, reviews experience and suggests options for reform for three economic instruments managed by MoLIT. It sketches ways to move Smart Water Management forward and identifies some of the bottlenecks of prevailing water allocation regimes, which need to be addressed.

Remarkable achievements, facing new challenges

Water management in Korea has been able to cope with and to facilitate rapid urbanisation and economic growth. To do so, it had to harness a difficult hydrology, marked by variable water availability.

Within one generation, Korea has transformed itself from a poor agrarian society to a modern industrial nation, a feat rarely seen before. Rapid economic development has gone hand in hand with fast urbanisation. Water supply augmentation has been a driving factor of industrial, agricultural and land development in Korea. Since 1965, the government has invested heavily in quantitative development strategies to meet water needs deriving from fast economic development and urbanisation (water use doubled between 1980 and 2007).

This was achieved despite complex climate characteristics that reveal significant seasonal and regional variations in water availability: the south and north-east parts (Gangwon) of the country experience over 1 400 mm of rainfall annually whereas the south-east region (Gyeongsang including the Central Nakdong River area) receives less than 1 100 mm of rainfall. Despite seemingly abundant precipitations, Korea is a thirsty country. As of 2015, Koreans abstracted 33% of total available water, putting the water balance at risk. Since the volume of total water use exceeds the amount of normal water runoff, which is measured during the off-flood season, the flood runoff needs to be reserved in impoundments.

A number of long-term trends currently observed in Korea are expected to affect water security and the capacity of the current water management system to adequately respond to water risks today and in the future: demographic trends, in particular rapid ageing of the population, will drive water demand and strain revenues; economic trends and fiscal consolidation will affect the capacity of public funds to finance water management; and climate change will have an impact on water availability and the exposure to water-related risks.

Policy responses based on infrastructure development and innovation

Korea has had to adapt its water system to manage variability in water availability and to provide even access to water supplies across the country. A sophisticated and extensive network of large, medium and small dams, irrigation facilities and multi-regional water systems was put in place over the course of 20 years (1980s to early 2000s) to supply water for domestic, industrial and agricultural uses, prevent floods and droughts, and generate hydro-electricity.

In 1961, the River Act laid the foundation for water resources policy and governance by setting a national framework for rivers. The Korea Water Resources Corporation Act, last amended in 2012, created K-water, the Korea Water Resources Corporation with a mandate to build, operate and manage all water resources development facilities in the country.

The Ministry of Land, Infrastructure and Transport (MoLIT) plays a central role in promoting water use efficiency in Korea. It is responsible for water quantity management and multi-regional water supply. It establishes policies for the sector, prepares plans for the sector's development, and oversees river basin management. It also exercises oversight authority for the institutions involved in carrying out the mission within its purview (including K-water).

State of play: Economic instruments to promote water use efficiency in Korea

MoLIT administers three economic instruments for water quantity management:

- A river water use fee is levied on water users (energy generators, industry, and domestic users) who abstract water from a river. It is levied by provinces and metropolitan cities and earmarked for the expenses on river maintenance and repair. The rate is set nationwide by the River Act Enforcement Decree. Local authorities are expected to pay the river water use fee; however, they do not report explicitly about revenues collected and expenditures covered.
- A dam water tariff is a uniform volumetric tariff levied on water users who abstract water from a river or a dam via a contract with K-water. The tariff is set nationwide to recover the cost of the construction and operation of dams used to store and supply water. Revenues are collected by K-water and earmarked for dam management.
- A multi-regional water tariff is a two-part tariff set nationwide to recover the costs of supplying water through multi-regional systems. It covers the cost of abstracting water (river water use fee or dam water tariff) and the cost of treating and supplying water to the user. Different tariffs apply to different water quality grades (raw water, settled water, purified water). Revenues that correspond to abstraction are transferred to the relevant authority. Additional revenues are earmarked for the operation of multi-regional water supply systems.

These instruments are too low to cover the costs of the infrastructure: tariffs for dam and multi-regional water supply barely cover operation and maintenance costs and partial capital charges; revenue covers only 50.3% of total expenditures as of 2013 in Korea. This situation contributes to ageing infrastructures and creates unnecessary water and expenditure waste. The percent of ageing infrastructures (over 20 years) in multi-regional supply systems is expected to reach 49.6% by 2020 and 79.6% by 2025 out of total facility capacity.

These instruments were not designed to promote water use efficiency or address scarcity. Where water is scarce and users compete to access the resource, water authorities are left with two options: either deprive new users from access to water, thus generating an opportunity cost and hindering economic development; or augment supply through additional infrastructures, generating additional public expenditures now and in the future.

This situation is compounded by water allocation regimes, which fail to allocate water where it is most useful. It is not consistent with the recent emphasis on water use efficiency and innovation, embodied in Smart Water Management in Korea – an effort to substantially enhance water use efficiency in Korea through information and communication technology (see below).

Economic instruments to promote water use efficiency in Korea

It would be beneficial to explore several avenues to make the best use of existing economic instruments and promote water use efficiency in Korea.

First, existing instruments could be adjusted to effectively contribute to water policy objectives:

- Revenues from river water use fee should be monitored and expenditure programmes financed from these revenues could be reviewed by MoLIT (at national or basin level). The river water use fee could then be revised based on explicit programmes of measures that contribute to river maintenance in the basin.

- Dam water tariff and multi-regional water tariff should reflect the costs of supply at the abstraction point. One way to do so is to ask government and service providers to set objectives for infrastructure development and service provision; to develop an investment plan to achieve these objectives; and to propose a financing strategy including tariffs) that cover the cost of the investment. The plan and the tariff would be reviewed by an economic regulator to check whether the investment is needed and the plan properly managed. Stakeholders would be consulted while investment programmes and appropriate financing strategies are developed.

Second, an abstraction charge could be considered in the longer term, which reflects the scarcity of water. An abstraction charge would operate as a signal to encourage water use where water is abundant and discourage it where water is scarce. Where or when water is abundant, the abstraction charge would be low; where and when water is scarce, the abstraction would be high, to reflect the opportunity cost of using water, when or where water is scarce, using water can deprive another user of the water he or she needs.

This report illustrates how such charges operate in several OECD countries (Belgium, Canada, Czech Republic, Estonia, France, Germany, Israel, the Netherlands, Portugal, the UK).

Several conditions should be met prior to implementing the proposed revenue structure:

- Multi-stakeholder buy-in as a result of a nationwide consultative process.
- Clear communication regarding the objective of the charge and transparent design of the charge and the use of collected revenues.
- The opportunity costs of using water and scarcity risks are clearly reflected in the charge.

Currently, investment in water infrastructure in Korea is set up to even out conditions of access to water across the country. Transition towards an abstraction charge would therefore need to take into account the current system. This report puts forth a gradual, long-term, inclusive, staged approach to introduce an abstraction charge, engaging stakeholders at both national and local levels. As an intermediary step a shadow abstraction charge could be introduced during the transition to a new system to raise awareness and promote an understanding of abstraction charges, as well as to gain experience with the tool before it is implemented.

Third, the institutional framework should be amended to promote water use efficiency in Korea and accompany the reform of economic instruments. The institutional framework should be able to reflect local conditions and engage with relevant stakeholders at the right scale (national, basin, or else). Some recent developments along these lines by MoLIT and K-water should be encouraged:

- Educate and inform the population and water users about water scarcity, the opportunity cost of mismanagement of water and the cost of supplying water.
- Develop a strategic programme towards stakeholder engagement, with result-oriented performance management.
- Mandate basin organisations under MoLIT's jurisdiction to engage with stakeholders at basin level.

Smart Water Management in Korea

Smart Water Management in Korea (K-SWM) seeks to establish a new paradigm for water resources management in a context marked by more uncertainty about future water availability and demand and increased risks of scarcity and floods. It has the potential to contribute substantially to water use efficiency in Korea and abroad.

K-SWM is designed to make the best use of information and communication technology to support (and integrate as much as possible) water resources management at different scales: from dam management and flood prevention, to detection of leakages and promotion of water use efficiency in homes. K-SWM was first rolled out in Korea as of 2015.

A combination of measures can expedite the deployment of K-SWM in Korea and abroad: 1) promote an integrated approach to water management (possibly through the Water Management Committee under the Prime Minister's Office); 2) gradually co-ordinate the deployment of smart water management to all water uses not under the responsibility of K-water; this would require extensive consultations with the administrations in charge, which are beyond the ambition of this policy dialogue; 3) reflect the full cost of supplying water in water supply and sanitation charges and tariffs to make profligate uses of water costly; 4) involve final water users in the definition of additional services that contribute to their needs; or 5) build capacities in municipalities.

Water allocation regimes: An important policy instrument to enhance water use efficiency

In principle, since the River Act in 1961, the right to use river water in Korea is granted by the government, via Flood Control Offices, and all water uses should be registered and permitted. Flood Control Offices review applications for new or additional water use. The four river basins are currently closed for new allocation, meaning that no river water is available for additional uses. As a consequence, new users need to apply for dam water. In the current context, dam water is available in every basin, and no revision of existing entitlements for addressing additional water demand has been required so far.

The coexistence of water entitlements acquired before and after the construction of dams creates legal disputes and inequities as regards who can access water and under which conditions. Senior right holders benefit from a rent situation and only use a portion of their entitlement, while new comers may be denied access to river water. The system is economically inefficient (it does not allocate water where it creates most value) and socially inequitable (risks of water scarcity are inequitably shared).

International experience indicates how welfare gains can derive from water allocation reforms without (or with only minimal) additional investment in water infrastructure. Korea would benefit from a systematic review of its water allocation regime, an ambition beyond the scope of this paper. OECD tools, drawing on international good practices, are available to help assess the performance of water allocation regimes and pave the way for reform.

Chapter 1

Water management in Korea at a turning point

This chapter characterises water resources in Korea. It reviews past experience with water management. The focus is on the management of water quantity, which is a prerogative of the Ministry of Land, Infrastructure and Transport.

The chapter also documents emerging trends, which challenge the prevailing model and call for innovative approaches to manage risks of water scarcity. It sets the scene for a discussion of challenges to be addressed and options to be considered.

The statistical data for Israel are supplied by and under the responsibility of the relevant Israeli authorities. The use of such data by the OECD is without prejudice to the status of the Golan Heights, East Jerusalem and Israeli settlements in the West Bank under the terms of international law.

Korea has been able to harness water resources to support rapid economic growth and urbanisation. This remarkable achievement materialised through supply augmentation and infrastructure development financed by the central budget. To cope with future challenges driven by rapid ageing of the population, a changing climate, and fiscal consolidation, the Korean model would benefit from a transition towards a system that places more attention on water demand management, enhances water use efficiency and allocates existing water resources where they create the most value for the Korean society.

1. Key achievements in water management in Korea

Water management policies in Korea have been able to cope with and facilitate rapid urbanisation and economic growth within a difficult water landscape marked by variable water availability.

1.1. Coping with rapid economic and urban growth

Within one generation, Korea has transformed itself from a poor agrarian society to a modern industrial nation, a feat rarely seen before. Korea's Gross Domestic Product (GDP) increased 31 000 times between 1953 and 2014 (from KRW 47.7 billion to KRW 1 485 trillion in 2014), at an annual growth of 7.3% over the period. Economic growth in 2014 was 3.3% (compare to an average 1.7% in OECD countries). A combination of intensive investments in infrastructure and human skills and of the rapid and pervasive application of information and communication technology (ICT) on productive sectors has been extraordinarily effective in growing the economy at impressive rates.

Rapid economic development has gone hand in hand with fast urbanisation. Since the modernisation of the Korean economy, the shares of urbanisation and industrialisation have been closely correlated (Figure 1.1). The share of urban population doubled since 1970, from 40.7% to 81.9% (OECD, 2014). As industrialisation has further advanced, more

Figure 1.1. **GDP and urbanisation share in Korea (1970-2010)**

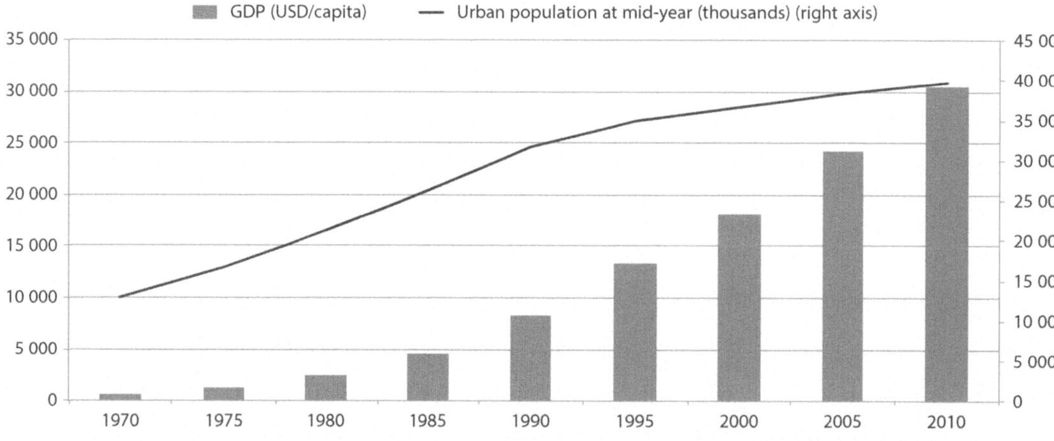

Note: The primary vertical axis (left hand) refers to GDP and the secondary vertical axis (right hand) refers to urbanisation share corresponding to the annual urban population at mid-year.

Source: Gross domestic product (GDP) indicator in OECD (2015), *National Accounts at a Glance*, http://dx.doi.org/10.1787/na-data-en; United Nations, Department of Economic and Social Affairs, Population Division (2014). World Urbanization Prospects: The 2014 Revision.

labour forces and capital have been absorbed into urban areas. Meanwhile, urban areas have provided the right scale of markets and large-scale infrastructure to create jobs and innovation capacities, adding economic growth potential (OECD, 2012a).

Water supply augmentation has been a driving factor of industrial, agricultural and land development in Korea. Since the first 10-year plan for water resources development in 1965 and the commission of the Seomjin-gang Dam, the first multi-purpose dam in 1965,[1] the government has invested heavily in quantitative development strategies to meet water needs deriving from fast economic development and urbanisation (water use doubled between 1980 and 2007, according to the 4th Long-term Comprehensive Plan for Water Resources, prepared by MoLIT). These strategies focused on the construction of large multi-purpose dams, reservoirs for agriculture water and the expansion of facility storage. Of the total amount of annual water abstraction in Korea, dam water account for 47.1% and river water for 44.7% (groundwater covers only 1.4%). Infrastructure development also achieved high rates of access to clean water and sanitation, with national averages of 98.6% and over 92.5% in 2014 respectively (OECD, 2012a). The highest levels of drinking water supply are registered in metropolitan cities (e.g. 100% in Seoul, Busan, Daegu and Daejeon, as of 2014).

1.2. Coping with seasonal and regional variability in water availability

Korea's water resources are administratively divided into six areas: Han River in the Seoul metropolitan area and Gyeonggi, Geum River in Chungcheong, Seomjin River and Yeongsan River in Jeolla, Nakdong River in Gyeongsang and Jeju/Ulleung islands (see Annex 1.A1 for further details).

Korea shows complex climate characteristics that reveal both seasonal and regional variations in water availability. It experiences yearly average precipitations of 1 274 mm (1.6 times the world average; see Annex 1.A1), more than half of which falls during a distinct rainy period (June to August), while the winter precipitation is less than 10% (Figure 1.2). The rainy season brings frequent flash floods. In addition, Korea's steep mountainous topography initiates high-runoffs and reduced opportunity for soil infiltration, which also contributes to flooding. Regional disparities in rainfall are marked: the southern and north-eastern parts (Gangwon) of the country experience over 1 400 mm

Figure 1.2. **Monthly rainfall in Korea (mm)**

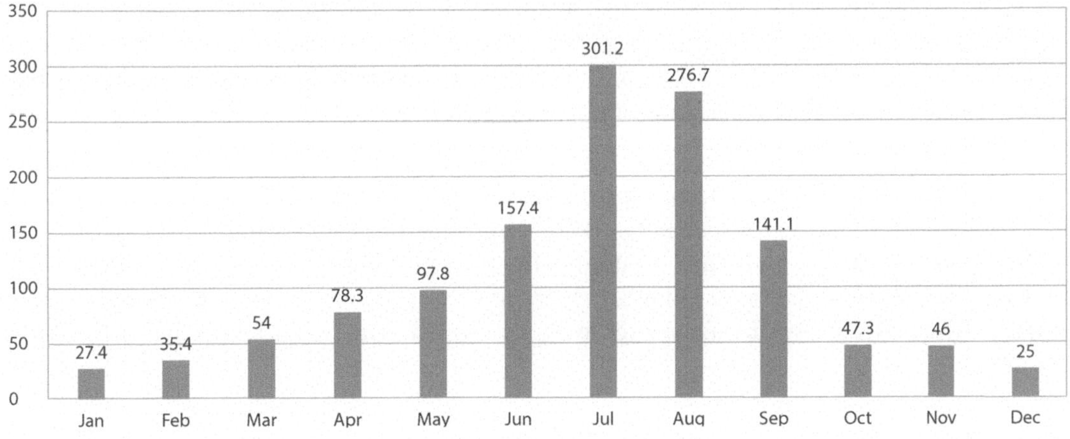

Source: OECD elaboration based on MoLIT (2017), *The 4th Long-term Comprehensive Plan of Water Resources (2001-2020)*, 3rd revision.

of rainfall annually, whereas the Southeast region (Gyeongsang including the so-called Central Nakdong River area) receives less than 1 100 mm of rainfall (see Annex 1.A1).

In spite of seemingly abundant precipitations, Korea has limited water supply. The OECD classifies Korea as a water-stressed country. It is rapidly depleting its water resources compared to some OECD countries. As of 2015, Koreans had abstracted 33% of total available water, putting their water balance at risk; while Canadians withdrew roughly 1.2% (OECD, 2012c).

Figure 1.3. **Water stress in OECD countries**

2009 or latest year available; water abstractions as a % of renewable resource

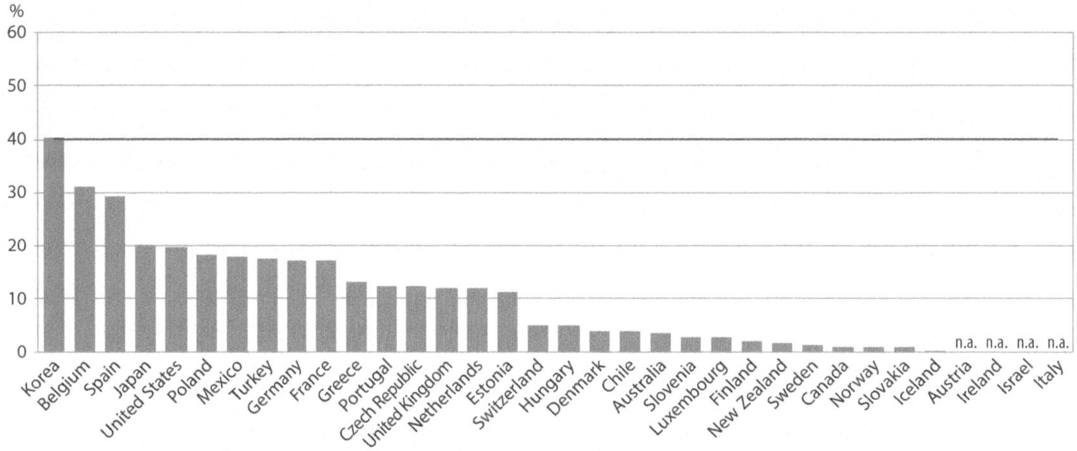

Note: Water stress: below 10% = no stress; 10-20% = low stress; 20-40% = medium stress; above 40%: severe stress.

Source: OECD (2012c), *OECD Environmental Outlook to 2050*, http://dx.doi.org/10.1787/9789264122246-en.

The total amount of water resources in Korea is 132.3 billion m^3 (see Figure 1.A1.8) with 76.0 billion m^3 (57%) available for use. In 2014, the total amount of water use was 37.2 billion m^3, accounting for 28% of the total water resources. Since the volume of total water use exceeds the amount of normal water runoff, which is measured during the off-flood season, the flood runoff needs to be reserved in impoundments. Among the total water use, the water use for household, industry and agriculture amounts to 25.1 billion m^3 per year or approximately 33% of the available water resources. The water use for household, industry, and agriculture are 7.6 billion m^3, 2.3 billion m^3 and 15.2 billion m^3 respectively. These figures suggest that water management in Korea, and in particular the management of scarcity risks, cannot ignore agriculture uses.

Korea has had to adapt its water system to manage this variability and to provide even access to water supplies across the country. This effort has materialised in the form of a sophisticated and extensive network of large, medium and small dams, irrigation facilities and multi-regional water systems aimed to provide water supply for domestic, industrial and agricultural uses, control river flows to prevent floods and droughts, and generate hydro-electricity. It was progressively developed from the 1980s to the early 2000s (Figure 1.4).

Figure 1.4. **Water abstraction and multi-purpose dams in Korea**

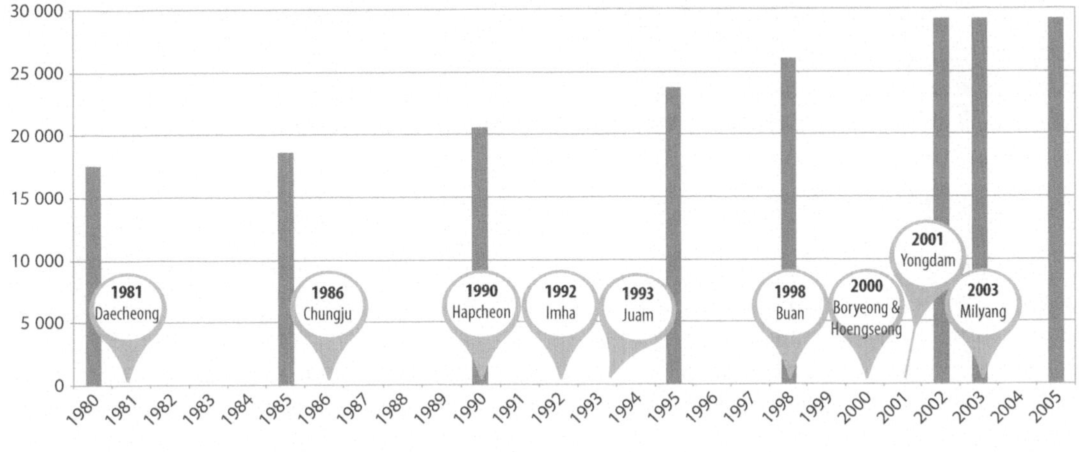

Note: The bars represent total water withdrawals (in billion m³). Grey circles mark the construction dates of some of Korea's main multi-purpose dams between 1980 and 2005.

Sources: K-water Water Resources Facilities, http://english.kwater.or.kr/eng/busi/resoPage.do?s_mid=1179.

Water management policy culminated with the Four Rivers Restoration (4RR) project, a large scale project spearheaded by former president Lee Myung-bak in 2009. The 4RR project was designed as a multi-purpose, green-growth infrastructure initiative with five key objectives: i) to secure water supply; ii) flood control; iii) improve water quality and restore ecosystems; iv) develop riparian space for cultural and leisure activities; and v) regional development centred on rivers. To achieve these core tasks, the 4RR project planned the construction of major facilities and infrastructure, and involved five ministries and 78 local authorities. The 4RR project was completed in 2012.

Table 1.1. **Details of the Four River Restoration Project**

Project period	October 2009-December 2012
Project area	Total length (km): 1 266 (Han River 255, Nakdong River 470, Geum River 272, Yeongsan River 269)
Project scope	Total 170 construction zones
Major facility	Dredging: 450 million m^2 Weir: 16 Bank reinforcement: 784 km New dam: 2 Reservoir connection: 1 Flood retention/storage reservoir: 5 Reservoir embankments: 110 Farmland remodelling projects: 140 Environmental treatment facilities: 1281 Flood retention reservoir: 2 Small-size hydro-power: 16; total capacity: 50 771 kW
Project cost	Total 19.4 billion USD MoLIT: 13.5 billion USD MAFRA: 2.6 billion USD ME: 3.3 billion USD

Source: Four MRR Project White Paper.

Through Four Rivers Restoration Project, the number of large floods decreased, more water resources have been secured and certain water quality indicators improved. At the same time, slowed river flow caused the population of some aquatic species to decrease and contributed to algae blooms in some areas; dredging and construction of riverside eco-parks affected some habitats and species. Continuous monitoring is therefore important to keep track of the long-term aquatic ecosystem and river beds variations.

2. Key future trends affecting water security in Korea

A number of long-term trends are expected to affect water security and the capacity of the current water management system to adequately respond to water risks today and in the future. These include demographic trends that will drive water demand and strain revenues; economic trends that will affect the capacity of public funds to finance water management; and climate change that will have an impact on water availability and the exposure to water-related risks.

2.1. Demographic trends

Two demographic challenges in Korea are expected to challenge prevailing approaches to water resources in the future. The primary challenge is related to the number of one-person households, followed by trends in ageing in the country.

While the demographic growth rate of Korea's population is expected to slow down, the number of one-person households has dramatically increased, as it has in other OECD countries. The share of single-occupancy households rose sharply, from 9% to 27.1%, over the period 1990-2015, as compared with the rise from 22% to 29% in OECD countries. One-person households are typically associated with higher per capita consumption of water.

Figure 1.5. **Percentage of one-person households in selected OECD countries**

One-person households divided by the total number of households – 1980 and 2008

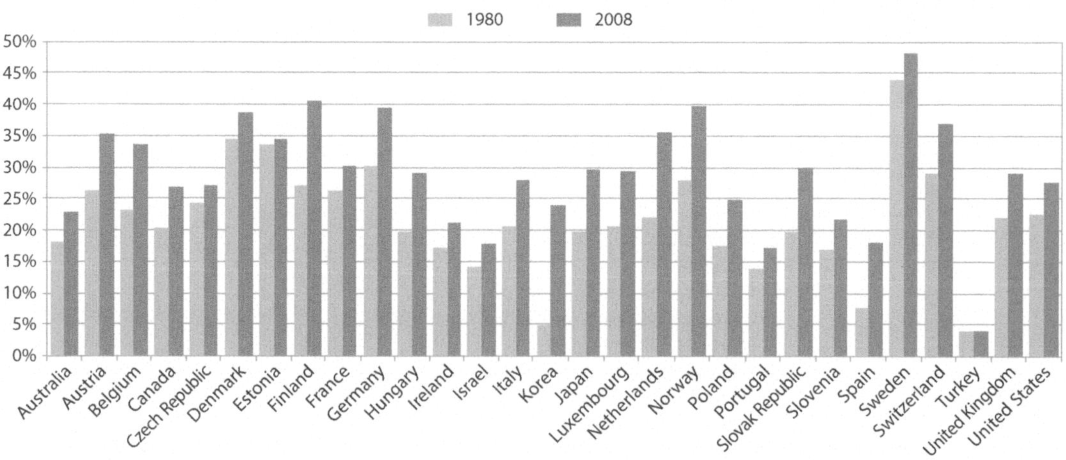

Source: OECD (2012d), *Compact City Policies: A Comparative Assessment*, http://dx.doi.org/10.1787/9789264167865-en.

In addition, the Korean population is ageing rapidly, transitioning from the fourth youngest OECD country today to the third oldest by 2050, behind Japan and Spain. It is projected to be the fastest population ageing in the OECD (OECD, 2016a). In 2012, the country's fertility rate was low at 1.2% compared to 1.7% on average in OECD countries. In 2050, more than one-third of the population will be over the age of 65. An increase in the percentage of elderly residents will decrease the tax base and strain revenues from the water sector as a special tariff discount applies to people over 65 under the Water Supply and Waterworks Installation Act (i.e. the recipients eligible to or of assistance under the National Basic Living Security Act, including elderly citizens over 65 years old, are entitled to a water tariff reduction). An ageing population also puts more demand on public budget (typically for health expenditures).

Figure 1.6. **Population ageing in Korea is projected to be the fastest in the OECD**

Population aged 65 and over as a percentage of the population aged 15 to 64

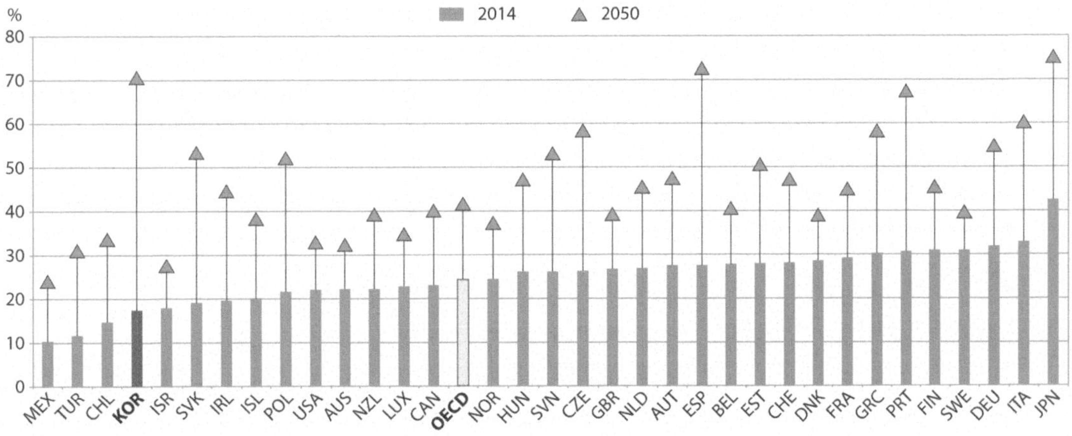

Source: OECD (2016a), *OECD Economic Surveys: Korea*, OECD Publishing, Paris. http://dx.doi.org/10.1787/eco_surveys-kor-2016-en.

2.2. Climate change

Average temperature has been rising gradually over the past 90 years in Korea, and sea levels have risen on average by 4.02 mm per year between 1993 and 2008, 30% higher than the global average of 3.16 mm per year. Precipitations show an increasing trend as well. Between 1996 and 2005, average precipitation increased by 10% and the number of days with over 80 mm of precipitation increased from 20 to 28 days. In addition, precipitation run-off has surged due to increasing impervious areas caused by urbanisation.

Water availability is becoming less predictable and the occurrence of extreme climate events, such as drought, high temperature and torrential rainfall is rising in frequency and intensity. Korea experienced the heaviest rainfall in 100 years in 2011, the worst drought for the first time in 100 years in 2012. The extreme drought in 2015 brought water supply restriction to eight municipalities in Chungcheongnam-do Province. More recently, the 2015 prolonged drought was the worst drought in a century in the Chungcheong area, and the third most severe drought throughout the country. These trends increase the potential for coastal erosion and severe flooding, thereby putting populations and infrastructure at risk.

Other projected climate change impacts on Korea were compiled by the OECD (OECD, 2013) and include increase of annual precipitation in six major cities over the next 100 years; two-fold increase in rainfall intensity, as compared to the 1970s; increase in the frequency of longer droughts and higher temperatures, which will generate algae blooms in water systems that make it more difficult to control the quality of tap water; and changes in the distribution of aquatic ecosystems caused by changes in water temperatures and volume.

Korea's flood risk index[2] is higher than other countries (see Figure 1.7). Over the last decade (2006-15), flood damages totalled KRW 4 899.5 billion (Han river basin: KRW 2 047.7 billion) and affected 196 915 people (Han river basin: 135 900) (Ministry of Public Safety and Security, 2015).

Figure 1.7. **Flood risk index by country**

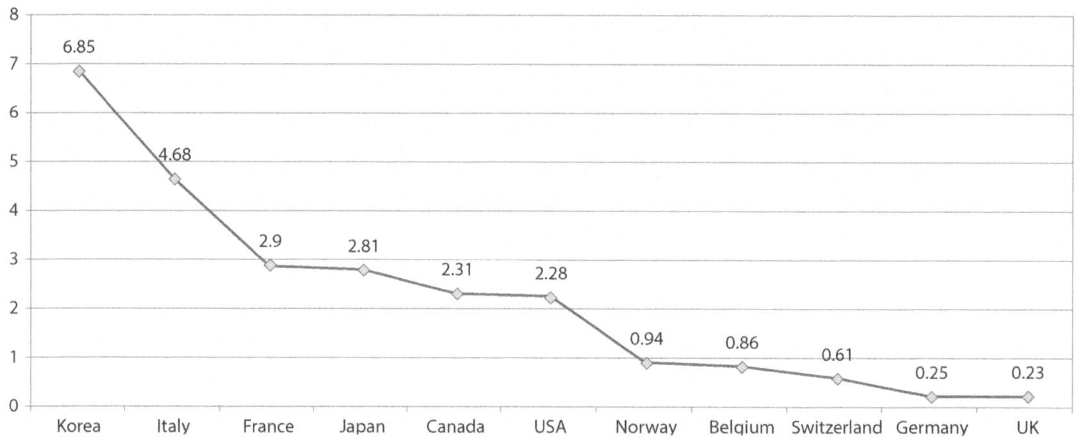

Source: based on UNDP (2004), *Reducing Disaster Risk. A Challenge for Development.*

Current policy measures and long-term planning take climate change into consideration. The *2009-2030 National Comprehensive Plan on Climate Change Adaptation* provides an integrated climate change adaptation plan for the country. The *2016-20 National Climate Change Adaptation Plan* was launched in December 2015. The Act on the Promotion and Support of Water Re-use mandates the installation of rainwater-equipped facilities and water reuse facilities in government buildings; this measure is intended to enhance water use efficiency and to reduce environmental impacts of new public buildings. Early warning systems have been established to minimise damage from disasters. Also, the Korea Adaptation Centre for Climate Change was created in 2009 to support the development of adaptation measures at central and local level. Research programmes are conducted on technological developments to reduce run-off and non-point source pollution, enhance the efficiency of waterworks and strengthen demand management for tap water.

2.3. Economic trends and fiscal consolidation

Korea is experiencing a period of slower growth. The *OECD Economic Survey of Korea* (OECD, 2016a) notes that growth has slowed from an average of 4.25% per year over 2001-11 to 2.75% since 2011. The slowdown in world trade since 2010 has been especially detrimental to Korea, as exports account for nearly 60% of total demand. Moreover, Korea's exports have faced strong competition from emerging economies, notably China, and with advanced economies in high-end markets. Meanwhile, domestic demand has been constrained by structural problems, such as high household debt, stagnant service sector productivity and struggling small and medium-sized enterprises. Nevertheless, the government has a plan to scale-up public spending for infrastructure, with investment via social overhead expenditure (SOC) expenditure expected to amount approximately KRW 110.3 trillion for the period 2014-18 (G20, 2015).[3]

There is a concern that a tighter fiscal environment could threaten the stability of funding for water management, particularly given Korea's heavy reliance on the central public budget to manage water. In 2013, expenditures in water management totalled KRW 17.9 trillion, of which only 50.3% were covered by revenues from the sector, leaving the remaining amount to be balanced out with public budget (K-water, 2015). Although the share of expenditures covered by water revenues has been gradually increasing since 2011, the sector is still at risk given the variability of the central government's policy priorities.

Annex 1.A1

Data on water resources availability in Korea

Table 1.A1.1. **Average rainfall and available water per capita in selected countries**

	Korea	Japan	China	France	UK	USA	World
Average rainfall (mm/year)	1 274	1 668	645	867	1 220	715	807
Average rainfall per capita (m^3/year)	2 660	4 932	4 607	7 794	4 736	22 560	16 427
Available water resources per capita (m^3/year)	1 553	3 232	2 130	3 326	2 429	10 075	8 372

Note: Average rainfall per capita (in m^3/year) in a given area = average annual precipitation/population

Source: Water Resource Management Information System (WAMIS).

Figure 1.A1.1. **Water resources status in Korea**

100 million m^3/year

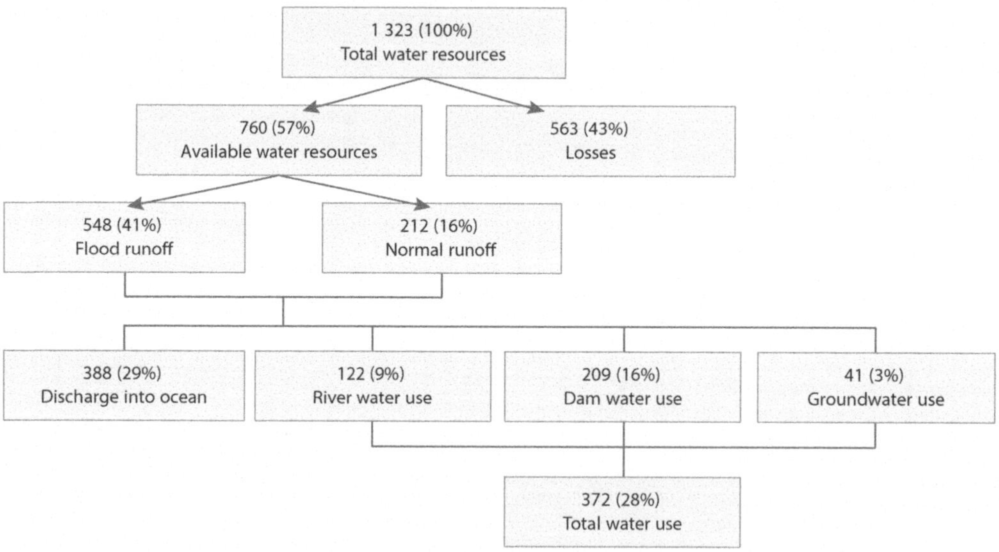

Note: Flood runoff is measured between June and September

Source: MoLIT (2017), *The 4th Long-term Comprehensive Plan of Water Resources (2001-2020)*, 3rd revision.

Table 1.A1.2. **Water abstraction rates**

Billion m³

	1980	1990	1994	1998	2003	2007	2014
Total Water Resources	114.0	126.7	126.7	127.5	124.0	129.7	132.3
Available Water Resources (A)	66.2	69.7	69.7	73.1	72.3	75.3	76.0
Total Water Use	15.3	24.9	30.1	33.1	33.7	33.3	37.2
Water use for household, industry and agriculture (B)	12.8	21.3	23.7	26.0	26.2	25.5	25.1
Water abstraction rates (%): (B/A)	19%	31%	34%	36%	36%	32%	33%

Source: MoLIT (2017), *The 4th Long-term Comprehensive Plan of Water Resources (2001-2020), 3rd revision.*

Table 1.A1.3. **Water stress indicator by abstraction rates**

Abstraction rates	Water stress level	Country examples
Below 10%	Low	New Zealand, Canada, Russia
10-20%	Middle	Japan, USA, UK, France, Turkey
20-40%	Mid-high	Korea, China, India, Italy, Singapore, South Africa, Spain
Above 40%	High	Iran, Iraq, Egypt

Source: MoLIT (2011), *The 4th Long-Term Comprehensive Plan of Water Resources.*

Figure 1.A1.2. **Average annual water availability in five inland hydrographic zones**

million m³

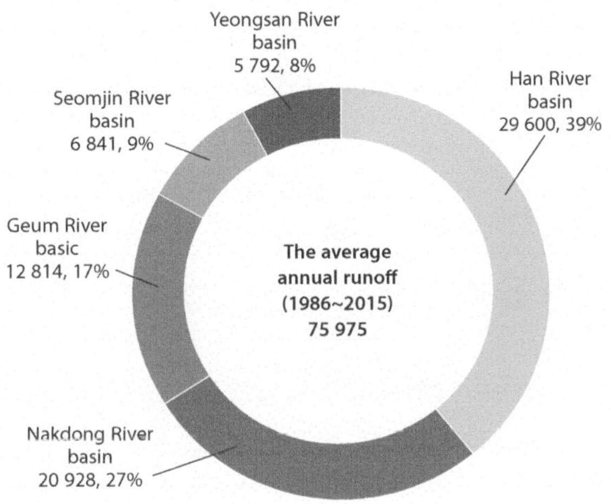

Source: MoLIT (2017), *The 4th Long-term Comprehensive Plan of Water Resources (2001-2020), 3rd revision.*

Table 1.A1.4. **Flow variation coefficients in rivers**

	South Korea			Other countries	
	Flow variation coefficients				
River	Before dam construction	1980-90	1995-2014	River (country)	Flow variation coefficients
Han	390	90	115	Oyodo (Japan)	110
Nakdong	372	260	101	Seine (France)	34
Geum	300	190	71	Nile (Egypt)	30
Seomjin	390	270	272	Rhine (Germany)	16
Yeongsan	320	130	214	Thames (UK)	8

Note: figures in parenthesis refer to coefficients before multi-purpose dams were constructed. Flow variation coefficients are defined as the ratio of the lowest daily flow to the largest daily flow in a year in the river

Source: MoLIT (2017), *The 4th Long-term Comprehensive Plan of Water Resources (2001-2020), 3rd revision.*

Table 1.A1.5. **Water use by purpose**

hundred million m^3

Quantity	1965	1980	1990	2003	2007	2014
Domestic	2	19	42	76	77	76
Industrial	4	7	24	26	28	23
Agriculture	45	102	147	160	154	152
Total	51	128	213	262	259	251

Source: MoLIT (2017), *The 4th Long-term Comprehensive Plan of Water Resources (2001-2020), 3rd revision.*

Table 1.A1.6. **Trends in average annual rainfall in Korea for the short term**

	2006	2007	2008	2009	2010	2011	2012	2013	2014	2015
Rainfall (mm)	1424.3	1450.1	988.7	1204.9	1444.9	1622.6	1479.1	1162.9	1173.5	948.2
% compared to average year	110%	112%	77%	93%	112%	126%	115%	90%	91%	74%

Source: Statistics Korea, www.index.go.kr/potal/main/EachDtlPageDetail.do?idx_cd=1401.

Annex 1.A2

A brief history of water resources development in Korea

In the 1960s, the Government of Korea started a comprehensive water resources development plan with river basin investigations. The government undertook full-scale development of water resources through the construction of large dams and flood-control projects in the 1970s and 1980s. In the 1990s, the government introduced eco-friendly water resources management, and in the 2000s it shifted the policy focus from water usage and flood control to sustainable management of water, including fluvial environment. Since 2010, the government has made efforts to cope with climate change and to upgrade water management systems.

Implementation of the Comprehensive Water Resources Development Plan (1960s)

As the nation officially began to develop its land in the 1960s, the Ministry of Construction was established and efforts were made to find new water resources. The River Act, a regulation to manage rivers mostly for controlling floods, has played the role of a mother law for river management and water resources development since it was enacted in 1961. The government established a 10-year plan for water resources development (1966-75) in 1965 and executed projects for flood control and for water usage at the same time. In 1966, the Specific Multi-Purpose Dams Act was legislated and the Korea Water Resources Development Corporation (now K-water) was established to implement the law.

A major project for investigating the basins of the nation's four major rivers was executed in the 1960s, in turn, a foundation for implementing a comprehensive and long-term development plan aimed at flood control and water usage, including the construction of multi-purpose dams, was established. Construction of the Seomjin-gang Dam, the nation's first multi-purpose dam, was complete in 1965. In addition, about 190 agricultural reservoirs were built and irrigation and reclamation projects were carried out to secure agricultural self-sufficiency. In addition, hydroelectric dams were built to secure electric power that served as a driving force for Korea's rapid industrialisation.

Implementation of the Comprehensive Water Resources Development in the 1970s

With the establishment of a plan for developing the basins of four major rivers (1971~81) in connection with the first plan for developing national land, consistent development programmes within the basin, including plans for constructing multi-purpose dams for controlling floods in the basins of those rivers, improving rivers and building irrigation systems and estuary dikes, were carried out. During this period, multi-purpose dams, such as the Soyang-gang Dam (1973) and Andong Dam (1977), were completed. Additionally, the first-phase of the multi-regional water supply system in the Seoul Metropolitan Area was built to provide stable water to meet the living and industrial needs.

Advancement of the Comprehensive Water Resources Development in the1980s

In conjunction with the rapid development of the Korean economy, water resources development also expanded rapidly with water consumption tripling from 5.1 billion m^3 in 1965 to 15.3 billion m^3 by 1980. After the establishment of the comprehensive water resources development plan, multi-purpose dams, such as the Chungju Dam (1986) and the Hapcheon Dam (1989), were built and efforts were made to construct multi-regional water supply systems in the Seoul Metropolitan Area and other regions. As well, 249 agricultural reservoirs were constructed to supply agricultural water stably. As the necessity to manage water quality emerged during this period, the Environmental Office was created and was responsible for managing the water quality of public waters. River maintenance commenced, which took into consideration water friendly functions, such as the Han River synthetic development project (1982-86).

Commencement of Eco-Friendly Water Resources Management in the 1990s

The government established a comprehensive long-term plan for water resources development (1991-2011) to satisfy the increasing demand for water in industrial and urban areas through the construction of mostly small and medium scale multi-purpose dams (10 dams with a total capacity of 3.7 billion m^3 and power generation capacity of 600 GWh) and expansion of multi-regional water supply systems (26 systems with a total capacity of 10 million m^3). The comprehensive long-term water resources development plan was amended (1997-2011) and incorporates the need to undertake eco-friendly development and management of water resources to reflect the importance of water quality management of rivers and reservoirs.

As water quality became a social issue after the phenol pollution incident in the Nakdong River (1991), the Environmental Office was promoted to the Ministry of Environment (1994). Water supply and drainage systems were transferred under the auspices of the Environment Ministry from the Ministry of Construction. After the River Act was completely revised in 1999, the basic plan for improving rivers included a plan to protect river environments by applying environment-friendly methods and the concept of in-stream flow was adopted.

Advancement of Eco-Friendly Water Resources Management in the 2000s

Due to ever-increasing national demands for improved water quality and environment, the paradigm of water resources development shifted during this period from supply-oriented quantitative development to the securement of water quality, ecosystems and environment of rivers. In 2001, the government established a comprehensive long-term plan for water resources development (Water Vision 2020) to strengthen policies regarding demand management and efficient use of water resources. The plan was amended in 2006 to create a water resources consultative body inclusive of related ministries, civil organisations and private experts and to build environment-friendly small and medium dams, like the Gunwi Dam.

In 2006, the amount of in-stream flow of 60 points of mainstreams and important tributary streams of the nation's five major rivers were notified for eco-friendly management of rivers. In 2008, the River Act was amended to secure the amount of in-stream flow as well as eco-friendly projects for improving national and provincial rivers were executed.

The government continued reform efforts to prevent damages from floods, securing water and protecting river environments. Two important projects were the Four Major Rivers Restoration Project (4MRRP) and the Gyeongin Ara Waterway Project.

The water supply ratio in farming and fishing areas increased following the expansion of multi-regional water supply system capacities. As of 2015, the nation's water supply penetration ratio was more than 97%. The government has also tried to improve the quality of water supply systems through the introduction of advanced water treatment facilities, as public demand for clean and healthy tap water increased.

Consideration of climate change adaptation in water management (2010s)

According to the second amendment (2011) of the comprehensive long-term plan for water resources development, the government established a plan for securing water to cope with climate change and a strategy to support the water industry.

Notes

1. A brief history of water development in Korea is appended.

2. The index refers to relative vulnerability defined by the UNDP as the numbers of casualties per million people exposed. See UNDP (2004), *Reducing Disaster Risk. A Challenge for Development*, UNDP (pages 144 and following).

3. In particular, the focus will be placed on constructing ports, industrial complexes, roads and railways. For example, the government will invest amount of KRW 2.1 trillion from 2015 to 2019 in building Great Train eXpress (GTX), a high speed railway network that will connect metropolitan areas.

References

G20 Turkey (2015), Adjusted Growth Strategy 2015 – Republic of Korea. Available online: http://g20.org.tr/wp-content/uploads/2015/11/Adjusted-Growth-Strategy-2015-Republic-of-Korea.pdf (accessed on 16 June 2016).

K-water (2015), Research on water policy for efficient water management.

OECD (2016a), *OECD Economic Surveys: Korea 2016*, OECD Publishing, Paris. http://dx.doi.org/10.1787/eco_surveys-kor-2016-en.

OECD (2016b), *Road Infrastructure, Inclusive Development and Traffic Safety in Korea*, OECD Publishing, Paris. http://dx.doi.org/10.1787/9789264255517-en.

OECD (2015), *National Accounts at a Glance*, http://dx.doi.org/10.1787/na-data-en.

OECD (2014), *Compact City Policies: Korea. Towards Sustainable and Inclusive Growth*, OECD Green Growth Studies, OECD Publishing, Paris. http://dx.doi.org/10.1787/9789264225503-en.

OECD (2013), *Water and Climate Change Adaptation*, OECD Publishing, Paris. http://dx.doi.org/10.1787/9789264200449-en.

OECD (2012a), *OECD Urban Policy Reviews, Korea*, OECD Urban Policy Reviews, OECD Publishing, Paris. http://dx.doi.org/10.1787/9789264174153-en.

OECD (2012b), *OECD Economic Surveys: Korea 2012*, OECD Publishing, Paris. http://dx.doi.org/10.1787/eco_surveys-kor-2012-en.

OECD (2012c), *OECD Environmental Outlook to 2050*, OECD Publishing, Paris. http://dx.doi.org/10.1787/9789264122246-en.

OECD (2012d), *Compact City Policies: A comparative assessment*, OECD publishing, Paris. http://dx.doi.org/10.1787/22229523.

Ministry of Environment (2014), *Statistics of Sewerage*, Korea.

MoLIT (2017), *The 4th Long-Term Comprehensive Plan of Water Resources, 3rd revision*, Korea.

MoLIT (2011), *The 4th Long-Term Comprehensive Plan of Water Resources* (국토교통부 수자원장기종합계획), Korea.

Statistics Korea (통계청), www.index.go.kr/potal/main/EachDtlPageDetail.do?idx_cd=1401, data on trends in average annual rainfall in Korea for the period 2006-15.

UN DESA (2014), *World Urbanization Prospects. The 2014 Revision*, United Nations.

UNDP (2004), *Reducing Disaster Risk. A Challenge for Development*, UNDP.

Water Resource Management Information System, www.wamis.go.kr/WATERPORTAL/sub_10_01.html, water-related statistics published by Han River Flood Control Office.

Chapter 2

Policy, legal and institutional framework for water management in Korea

This chapter presents the policy, legal and institutional features that make up Korea's water management framework. It documents the main water policy trends that have prevailed in the last fifty years and chronicles key reforms that have marked water management in Korea. In addition, it discusses current reflections towards the adoption of a Basic Law on Water Resources Management.

While the focus is on policy instruments under the responsibility of MoLIT – namely economic instruments, smart water management (SWM) and water allocation – these do not operate in an institutional vacuum: they are embedded in a web of institutions. This chapter provides an institutional mapping of the multiple authorities and non-state actors involved in water management at central, basin and local level. It points at the strengths and limitations of water governance that affect how economic instruments, SWM and water allocation operate in practice.

The statistical data for Israel are supplied by and under the responsibility of the relevant Israeli authorities. The use of such data by the OECD is without prejudice to the status of the Golan Heights, East Jerusalem and Israeli settlements in the West Bank under the terms of international law.

1. Recent developments in policy and legal framework for water management in Korea

1.1. Dynamics of water policy

Until the mid-1950s, Korea was essentially a rural community involved in farming, predominantly paddy rice cultivation. This legacy, along with more recent evolutions in the country's economy and urban development (see Chapter 1) has shaped current water management institutions, policies and practices. Furthermore, major policy decisions and instruments for water developed in the 1970-80s were dedicated to supply augmentation that would enable economic growth and meet increasing water demand in cities.

Between 1980 and 2000, important water policy choices and changes have been largely influenced by the stresses of industrial development and environmental crises (Kim, Koh and al., 2007). Historically, Korea has been highly dependent on its rivers, in particular the four major rivers, as the primary source of drinking water. Coping with water quality concerns has been a chronic issue over the years given the lack of infrastructure in place to deal with major sources of pollution, e.g. wastewater. Sewage treatment plants barely existed in Korea prior to 1989, and in the 1980s, problems associated with industrial and domestic wastewater became national issues, as accidents occurred and environmental conditions deteriorated. In response, the government initiated six major programmes for water quality restoration, including the establishment of wastewater treatment plans and the first-ever nationwide Comprehensive Measure for Clean Water Provision (ADB, 2014).

At that time, the Environmental Agency, which set up the comprehensive measure, was also upgraded to the level of Ministry of Environment (ME). The new ministry expanded its capacities with increased financial support and functions transferred from the MoLIT (then called the Ministry of Construction and Transport) in 1994.

1.2. Main water reforms

Throughout the second half of the 20th century, water policy evolved towards integrated river management, planning and decentralisation in Korea. Water-related legislations were developed in a piecemeal fashion: in 50 years, more than 25 acts were enacted and revised by different ministries according to transitions in water policy (Table 2.1).

Table 2.1. **Korean laws related to water resources**

Act or Special measures	First enacted	Line Ministry
River Act	1961	MoLIT
Water Supply and Waterworks Installation Act	1961	ME
Public Waters Management Act	1961	Ministry of Oceans and Fisheries
Public Waters Reclamation Act	1962	Ministry of Oceans and Fisheries
Management of Drinking Water Act	1965	ME
Sewerage Act	1966	ME
Countermeasures against Natural Disasters Act	1967	MoPSS
Korea Water Resources Corporation Act	1987	MoLIT
Water Quality Preservation Act	1990	ME
Framework Act on Environmental Policy	1991	ME
Groundwater Act	1993	MoLIT
Environmental Impact Assessment Act	1993	ME

Table 2.1. **Korean laws related to water resources** *(continued)*

Act or Special measures	First enacted	Line Ministry
Small River Maintenance Act	1995	MoPSS
Drinking Water Management Act	1995	ME
Agriculture and Fishery Improvement Act	1995	MAFRA
Act on Dam Construction and Support for Surrounding Areas	1999	MoLIT
Act on Han River Water Quality Improvement and Community Support	1999	ME
Act on Geum River Watershed Management and Community Support	2002	ME
Act on Nakdong River Watershed Management and Community Support	2002	ME
Act on Yeongsan and Sumjin River Watershed Management and Community Support	2002	ME
National Land Planning and Utilisation Act	2003	MoLIT
Framework Act on the National Land	2003	MoLIT
Framework Act on the Management of Disasters and Safety	2004	MoPSS
Public Waters Management and Reclamation Act [*The act integrates the Public Waters Management Act and the Public Waters Reclamation Act*]	2010	Ministry of Oceans and Fisheries
Urban Development Act	2016	MoLIT

Sources: Ministry of Government Legislation, *Korean laws in English*, www.moleg.go.kr/english/korLawEng; OECD (2017), *OECD Environmental Performance Reviews: Korea.*

A number of key laws have come to form the backbone of water management in the country. In 1961, the enactment of the River Act – last amended in 2016 – laid the foundations for water resources policy and governance by setting a national framework for river management (see details on the objectives and scope in Box 2.1.). The Water Supply and Waterworks Installation Act adopted in 1961, and last revisited in 2016, marked the entrustment of drinking water supply to local authorities.

Box 2.1. **Korea's 1961 River Act: Objectives and scope**

The River Act was adopted on 30 December 1961 and defined the basic principles for water resources management in terms of the designation, use, and conservation of rivers. The River Act was last amended in 2016. The Act makes the distinction between *national* rivers (defined as "having vital importance to national land conservation or national economy") and *local* rivers (designated as "having close relation with local public interests"). The River Act established that the former category would be under the purview of MoLIT and the later under the jurisdiction of provinces and metropolitan cities.

The Act set the building blocks of a national framework for river management in terms of: i) *Policy coherence*, with the preparation of water control plans and survey that should take account of urban planning and land acquisition issues (Article 21); ii) *Information sharing*, with the creation of an information system on water resources by MoLIT (Article 22); iii) *Revenues*, with the introduction of fees for river land occupancy (Article 37) and river water use (Article 50); iv) *Institutions* and *stakeholder engagement*, with the establishment of the Central River Management Committee, under MoLIT, and local river management committees, under provinces and metropolitan cities, to deliberate on matter of river management and disputes. In addition to national and local government representatives these River Management Committees would include experts and academics (Article 87).

Box 2.1. Korea's 1961 River Act: Objectives and scope *(continued)*

The purposes of the River Act were to increase the benefits deriving from river use and preservation, nature-friendly river maintenance and the prevention of river-related damages, and to contribute to public welfare.

Source: Ministry of Government Legislation, *Korean laws in English*, www.moleg.go.kr/english/korLawEng?pstSeq=52772.

The 1987 Korea Water Resources Corporation Act marked another turning point with the creation of K-water and its mandate to build, operate and manage water resources development facilities (Article 9 [1]) and to collect related charges and fees. K-water currently manages 20 multi-purpose dams and one estuary weir (Nakdong River). It also manages 3 flood control dams and 14 water supply dams. In addition, K-water manages 48 multi-regional water facilities, with a total water supply capacity of 17.6 million m^3/day.

There have been several attempts in the last 20 years to establish a Basic Law on Water Resources Management to solve the legal segmentation problem, the most recent having been proposed in 2016. Previous drafts of the water law suggested different scenarios to integrate water management functions and systems such as the creation of a water co-ordination committee under the Prime Minister's Office or under the President's Office.

The draft law currently being examined by the General Assembly is an attempt to enact an overarching legislation on water management. It is expected to provide the foundations for water management in the country and to set overall principles for the management of water quantity, quality and water supply and sanitation. These include integrated water resources management, water rights, information sharing, water financing (i.e. user-pays principle) and stakeholder engagement. The draft law is considering the establishment of a National Water Management Committee that would be responsible for developing the National Water Management Master Plan and evaluating water management holistically with the main purpose of encouraging co-ordination across ministries and regions, and aligning priorities. In addition, Regional Water Management Committees would be set-up and endowed with similar responsibilities at subnational level.

Adopting such a Law on Water could have the potential to foster consistency and integration across the current water acts and would provide an incentive to co-operate across ministries. On-going discussions on a national water law provide a timely opportunity to clarify the national standpoint on water policy, manage water more effectively and cope with future challenges. The experience of the Netherlands can be useful, in this context (see Box 2.2).

Box 2.2. Setting-up overarching water legislation: The experience of the Dutch Water Act

Before 2009, Dutch water legislation was extremely fragmented. Over the years separate laws have been drawn up for every part of water management, which have their own evaluation framework, legal instruments, procedures and systems of appeal. This fragmented, reactive method of policy development was caused by the drafting of new laws to react to "disaster"

Box 2.2. **Setting-up overarching water legislation: The experience of the Dutch Water Act** *(continued)*

situations such as prolonged drought, imminent flooding, etc. As a result, the body of legislation did not respond to water management in a cohesive manner.

To address these challenges, the Water Act was adopted in 2009 and has been the quintessential integration of national water management legislation to date. The Water Act unified the piecemeal legislation by combining eight previous laws, and providing policy instruments for integrated water management.

The Water Act highlights a "water system approach" based on all relationships within water systems including the relationship between the quality and quantity of water, between surface water and groundwater, but also the relationship between water, land use and water users. The legislation regulates formal decisions on target water levels, and includes notification and permit requirement for water discharges and abstraction.

Sources: Dutch Water Authorities (2015), *Water Governance: The Dutch Water Authority Model*; OECD (2014), *Water Governance in the Netherlands: Fit for the Future?*.

2. Who does what at which level in water management

Korea's institutional framework for water management is multi-layered and multifaceted to accommodate the different water uses in the country: drinking water, irrigation, hydropower and environmental needs. The number of institutions, agencies, and other bodies involved in water management at central, basin and local levels attests of a fragmented institutional landscape (Figure 2.1). While these institutions allowed for fast and effective policies, they raise important co-ordination challenges for the development and implementation of integrated, coherent and inclusive water policy.[2]

2.1. Roles and responsibilities at central level

This section provides a picture of the allocation of roles and responsibilities for water policy design, financing, regulation and implementation at central level.

- The **Ministry of Land, Infrastructure and Transport** (MoLIT) is responsible for water quantity management and multi-regional water supply. It establishes policies for the sector, prepares plans for the sector's development, and oversees river basin management. It also exercises oversight authority for the institutions involved in carrying out the mission within its purview (including K-water, the public corporation responsible for water supply).
- The **Ministry of Environment** (ME) is responsible for local waterworks management, drinking water management and sewage policy. It makes policies, establishes laws and regulations, and implements necessary measures in relation to these issues. ME supervises related organisations, including KECO, a public organisation that supports water, sewage and water quality management.
- The **Ministry of Agriculture, Food and Rural Affairs** (MAFRA) is in charge of managing affairs related to securing stable food production, such as establishing and implementing policies for agricultural water development and management and supporting budgets and operating systems. Also, the ministry supervises Korea Rural Community Corporation – KRC's management of agricultural water.

Figure 2.1. **Institutional mapping for water resources management**

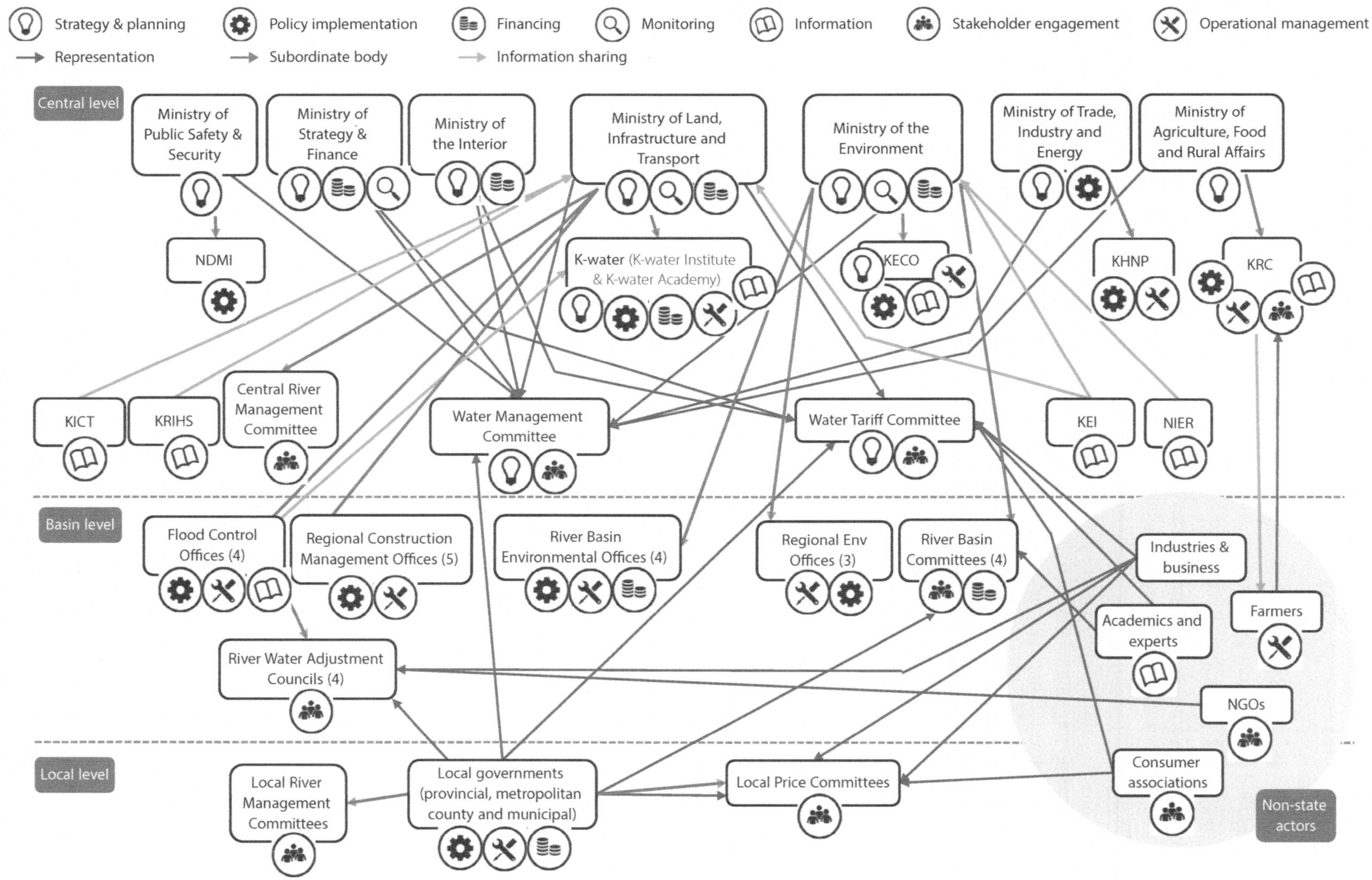

Note: KECO: Korea Environmental Corporation; KEI: Korea Environment Institute; KHNP: Korean Hydro and Nuclear Power Corporation; KICT: Korea Institute of civil Engineering; KRC: Korea Rural Community Corporation; KRIHS: Korea Research Institute for Human Settlements; KWI: Korea Waterworks Management Institute; NDMI: National Disaster Management Institute; NIER: National Institute of Environment Research

Source: Author's elaboration.

- The **Ministry of Strategy and Finance** (MoSF) is responsible for formulating and co-ordinating economic and fiscal policies, monitoring and reviewing public expenditures, and developing policies in regards to taxes, tariffs and public fund management, including for the water sector.
- The **Ministry of the Interior** (MoI) oversees decentralisation and the interactions between central and local authorities. It contributes to planning the tariff strategy for drinking water and sanitation service provision.
- The **Ministry of Trade, Industry and Energy** (MoTIE) assumes the responsibility for developing the national power development plan, which includes hydropower. It is also involved in the production of hydropower.
- The **Ministry of Public Safety and Security** (MPSS) assumes responsibilities for public safety and disaster prevention. It prepares the national disaster prevention plan.
- The **National Disaster Management Research Institute** (NDMI) works under the Ministry of Public Safety and Security to produce and disseminate practical disaster management guidance. It establishes the National Disaster Management Information Analysis and Sharing System and provides capacity building regarding disaster safety management technology development.
- **The National Institute of Environmental Research** (NIER) works under the Ministry of Environment and carries out research in water-environmental fields like total water pollution load management, water quality standard, water-environmental engineering, aquatic ecosystem and integrated water quality management and evaluation processes (e.g. water quality predictions, etc.).
- The **Water Management Committee** is a co-ordinating and advisory body gathering water-related stakeholders across levels of government. It was created by the 64th Prime Ministerial Directive of 4 November 2015 and cannot exceed 15 members who currently are the vice-ministers that play a role in water resources management, under the chairmanship of the Minister for Government Policy Coordination. Depending on the Committee's agenda, general-directors of state-owned corporations and agencies as well as and representatives of provinces and metropolitan cities are invited to participate. According to Article 6 of the Prime Ministerial Directive, the Committee is supported by a working-level committee with director-level public officials from central, provincial and metropolitan agencies involved in water management.
- The **River Management Committees (RMCs)**, set up under Article 87 of River Act, deliberate on important issues on river management and mediate disputes over the use of river water. Central river management committees exist under the Ministry of Land, Infrastructure and Transport. Local river management committees sit under local governments. They examine and mediate disputes over the use of river water, and deliberate on the following matters concerning national rivers: the designation or change of rivers or the revocation of designation, environmentally-friendly maintenance and conservation of rivers, and other matters relating to river management. The Central RMC can have up to 50 members, and each Local RMC up to 30. The Chair and the members of the Central RMC are appointed by MoLIT; the Chair is a staff of MoLIT and members can be academics, lawyers, or experts. The Chair and the members of the Local RMC are appointed by the City Mayor or Do Governor, and members belong to the same categories.

- The **Water Tariff Committee** is a deliberative body which discusses matters relating to water charges (for dam water and multi-regional water). It is chaired by the Director General for Water Resources Policy of the Ministry of Land, Infrastructure and Transport and comprises 15 members including the Ministry of the Interior's Director of State Enterprises, K-water's Director General of Management Services, experts and accounting specialists, selected representatives of water users (industries, local authorities), academics and consumer groups.

Four state-owned corporations act as operational agencies under the authority of ministries:

- The **Korean Water Resources Corporation** (K-water) is a public corporation acting under the authority of MoLIT and is responsible for the operation and management of water resources facilities. More specifically, it supplies bulk water to municipalities and industries through dams and multi-regional water supply systems. In some cases, K-water acts as service provider through a consignment contract with local authorities to manage water services.
- The **Korea Environment Corporation** (KECO) acts under the authority of ME, supporting policy making and implementation for water, sewage and water quality management. It is responsible for the operation and management of local waterworks through consignment contracts with local governments. It supports the establishment of National Waterworks Information System, the operation of Waterworks Information System, and the infrastructure for waterworks, sewage and water quality management.
- The **Korea Rural Community Corporation** (KRC) acts under the auspices of MAFRA. According to the Rearrangement of Agricultural and Fishing Villages Act and the Korea Rural Community Corporation and Farmland Management Fund Act, KRC is in charge of implementing development projects in farming and fishing village, and operating agricultural water and agricultural irrigation facilities within its management areas.
- **Korea Hydro- and Nuclear Power** (KHNP) operates ten hydropower plants under the authority of MoTIE.

Three research institutions produce the data, analysis and knowledge base to inform ministries:

- **Korea Environment Institute** (KEI) is a public research institute of the National Research Council for Economics, Humanities and Social Sciences under the Office of the Prime Minister. It provides science-based policy advice for government, public sector and civil society, including on water quality. KEI also reviews the statement of environmental impact assessment on the major development projects.
- **Korea Institute of Civil Engineering and Building Technology** (KICT) is a R&D think tank that focuses on construction policies and techniques and develops technology to improve public safety and the quality of life. In particular, KICT's Hydro Science and Engineering Research Institute work informs government policies and strategic plans regarding water resources, flood damage and drought control, waterway surveys, subterranean water mapping and climate change. The Hydro Science and Engineering Research Institute also provides hydraulic analysis and tests for rivers, estuaries, and coastal areas as well as evaluates and improves eco-friendly ecological engineering technologies, hydraulic structures, coastal structures and port infrastructure.

- **Korea Research Institute for Human Settlements** (KRIHS) is a research centre of the National Research Council for Economics, Humanities and Social Sciences, in the field of spatial policy, territorial development and competitiveness, and green growth. As part of its work on territorial planning, KRIHS focuses on policy research concerning water resources and river basin management.

Institutional fragmentation has led to policy inefficiencies in Korea. In the past, some decisions taken by one ministry have worked against decisions taken by another ministry. For instance, water quality standards, stricter discharge standards and expanded land use regulation in water source protection areas were established by ME in the early 1990s, but coincided with the deregulation of zoning policy applied by MoLIT (Kim, Koh and al., 2007). Another example is river management, which requires the consideration of sufficient water quantity, flood control and the environmental functions of rivers, but is tasked under distinct jurisdictions that do not interact regularly to devise co-ordinated solutions. Institutional fragmentation also leads to information asymmetries, whereby data and knowledge is scattered across many institutions thus making data sharing and comparison difficult.

Sectoral perspectives prevail in water planning. Several strategic plans are used in Korea as policy instruments. While their design is subject to some consultation across ministries, their roll-out remains compartmentalised, with different timelines (Table 2.2). Formulating a water master plan that would bring together the strategies of the different ministries in a concerted way would bring about benefits in terms of policy alignment (i.e. by harmonising the aims, contents and tools of these plans), financial efficiency (by exploiting synergies in water-related investments), and long-term policy-making (i.e. by developing a common vision for the sector).

Table 2.2. **Water-related strategic plans**

Plans	Line institutions	Duration
4th Long-term Comprehensive Plan for Water Resources	MoLIT	2010-20
Water Environment Management Master Plan II	ME	2016-25
Plan for Rationalising the Use of Water for Farming and Fishing	MAFRA	2015-24
3rd National Safety Management Master Plan	MPSS	2015-19
Energy Master Plan	MOTIE	2014-35

There have been several attempts to foster horizontal co-ordination regarding water management in Korea, but these have been fit for a specific and/or time-bound purpose and overall have fallen short in addressing co-ordination challenges in a structural way. These co-ordination platforms were often created to overcome a crisis. For instance, in Korea the Consultative Committee for Water Policy Adjustment was established in January 1997 (by Prime Minister's Instruction No. 344) following the phenol contamination of the Nakdong River from an industrial complex in Gumi. The Committee was considered the top decision-making organisation for the national water policy and brought together ministries with water-related mandates. In addition, the Water Quality Improvement Task Force was set up by presidential instruction No. 64 to support the Committee's activities and to help adjust water resources and water quality improvement policies of each ministry in question. Both the Committee and the Task Force were disbanded in 2005.

The recently established Water Management Committee appears as a step in the right direction to foster long-term ministerial co-ordination and policy alignment. This Committee was established on 4 November 2015, following the severe drought that occurred that same year, and has emerged as the main national body for resolving inter-ministerial issues. The Committee has supported effective co-ordination in times of droughts; however, it is not a permanent organisation based on statutory footing.

2.2. Roles and responsibilities at subnational level

By law, local authorities have primary responsibility for water supply and sanitation along with various levels of resources and capacity, which results in heterogeneity across the country.

- **Local governments** – they encompass provinces, metropolitan cities, counties and municipalities (Box 2.3.) – and manage local waterworks to supply domestic and industrial water either by directly abstracting water from their own water sources (e.g. local rivers) or by receiving bulk water from dams or multi-regional water supply systems. Most local governments with agricultural lands supply water to farmers. In addition, provinces and metropolitan authorities are responsible for collecting revenues from the river water use fee according to the River Act.

Box 2.3. **Subnational administrative units in Korea**

As a unitary state, Korea has a two-tier system of local government:

- The upper (or regional) tier (Territorial Level 3 in the OECD typology) includes Seoul Special City, which has the status of a capital city, six metropolitan cities and nine provinces (*do*).
- The lower (or basic local) tier (TL4 in OECD typology) is composed of 230 bodies, including 75 cities (*si*), 82 counties (*gun*, rural areas) and 69 autonomous districts (*gu*, urban areas that exist only in the metropolitan cities and Seoul). The lower tier of government is further divided into 3 477 administrative sub-branches, which are not legal entities and have no autonomous power: 215 *eup* (defined as the urban division of a county, or *gun*), 1 201 *myeon* (the rural division of a county, or *gun*) and 2 061 *dong* (which belong to cities including Seoul City, metropolitan cities and lower tier cities) (TL5).

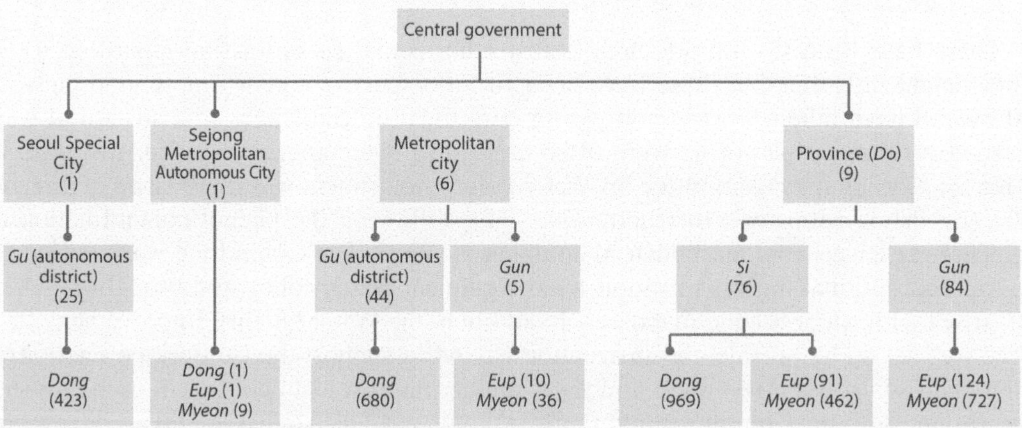

Box 2.3. **Subnational administrative units in Korea** *(continued)*

Within this institutional framework, the Local Autonomous Act designates the following units as urban: i) *si*, a lower administrative unit at the 4 level, with a population greater than 50 000; ii) *gu*, autonomous districts in metropolitan cities and in Seoul. Rural areas, known as counties or *gun*, are further divided into two categories: i) *myeon,* a basic subdivision of a gun; and ii) *eup*, an urbanised area in a rural unit, with a population of more than 20 000 people.

Source: OECD (2012a), *OECD Urban Policy Reviews, Korea 2012*, OECD Publishing, Paris. http://dx.doi.org/10.1787/9789264174153-en.

- **Local price committees** in each local authority deliberate on local prices, including for local water and wastewater services. It is composed of local government officials, local council members, consumer representatives, citizen representatives, academics, judicial officers and representatives from the media. Final decisions regarding water pricing are made by the local councils.

Various offices and committees were established at the watershed level to embody integrated water management:

- Five **Regional Construction Management Offices** function as "windows" of MoLIT as the subnational level. They were established by Presidential Decree (No.7663) with a broad mandate covering land management. They act as policy implementing bodies in charge of safety facilities, flood damage prevention, waterfront zoning and management, water control (i.e. preparation of basic river basin plans), river environment development, and flood simulation training, with consideration for land use policies and practices.
- Four **Flood Control Offices** work under the responsibility of MoLIT and are primarily data-acquiring/processing bodies. They are in charge of flood forecast, hydrological observation and hydrological data management for their respective basin. Based on the data collected from their monitoring stations, the Offices calculate forecast points which are then relayed to central agencies (MoLIT, K-water and the National Disaster Management Institute). The Offices are typically structured around a Forecast and Control Division that deals with dam operation and management, flood forecasting and river use permission (licencing), a Telecommunication Section that provides technological support to flood forecasting, and a River Information Centre that manages data acquisitions and analysis, operates the Water Resources Library and organises Flood Control Training for public officers. Flood Control Offices also issue water-use permits.
- Four **River Water Adjustment Councils** are hosted respectively by the Han, Nakdong, Geum and Youngsan Flood Control Offices, and mediate over water allocation issues. They were established in 2015 and 2016,[3] following tensions that arose during the 2015 drought and which required mediation across water users and stakeholders. Each council is chaired by the Director of the Flood Control Office and composed of water users with a water-use permit in the river basin, stakeholders, experts from the private sector and NGOs. It is in charge of co-ordinating river water allocation and deliberation over allocation conflicts when needed.

- **Local River Management Committees** are set up by provinces and metropolitan cities to deliberate on matter of river management and disputes.
- Four **River Basin Environmental Offices** were established for each of the major river under the jurisdiction of ME. They are in charge of river basin management, waste management and regional environmental impact assessments, as well as river basin management, pollution sources management and regional environmental impact assessments. They are also responsible for the approval of the Basic Plans for Sewerage Improvement, the maintenance of the drinking water protection areas and the water treatment facilities. In addition, they review, approve and evaluate the total pollution management system.
- Three **Regional Environmental Offices** are located in the cities of Wonju and Daegu and near the Saemangeum estuary. Under the responsibility of ME, offices approve basic plans for sewerage improvement, manage drinking water protection areas and the status of water purification facilities. They also review, approve and evaluate the "Total Water Pollution Load" management system. Offices are performing the basic functions of preserving water quality, managing water system, preserving natural environment, managing designated waste and chemical substances, discussing environmental effects evaluation, environmental monitoring and inspection. Their tasks include the establishment of water environment management plans for large areas and basin sewerage plan. They support the operation of water system committee.
- Four **River Basin Management Committees** (one for each major river) are established under the jurisdiction of the ME. They involve multiple stakeholders: they are chaired by the Vice-Minister of Environment and include MoLIT's Assistant Minister, vice-governors in each river basin and the CEO of K-water. River Basin Management Committees manage the river management funds (raised by from water use charges).

Figure 2.2. **The establishment of river basin organisations under MoLIT authority**

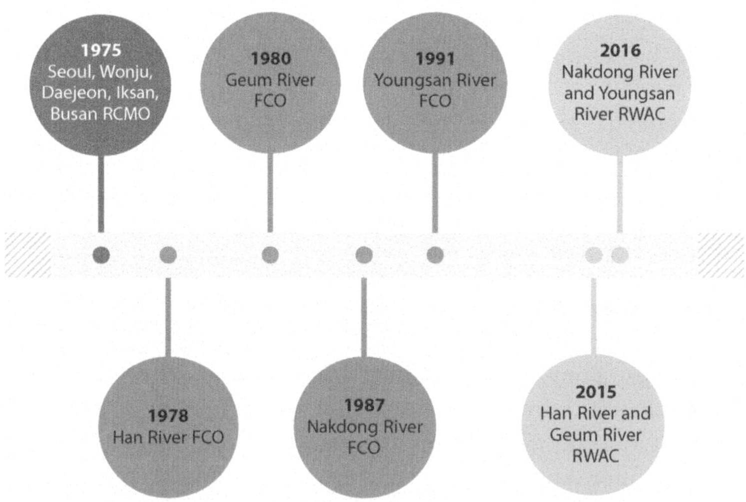

Note: RCMO: Regional Construction Management Office; FCO: Flood Control Office; RWAC: River Water Adjustment Council.

Basin institutions mostly mirror the dichotomy observed at national level between water quality and quantity management. Their strategies are defined separately by ME and MoLIT. River basin organisations under MoLIT were created in an incremental way. Figure 2.2 illustrates the staged establishment of Regional Construction Management Offices, Flood Control Offices and River Water Adjustment Councils over the spam of 40 years. Interestingly, river basin organisations under the responsibility of MoLIT were established according to two different water management logics: an administrative one for Regional Construction Management Offices that were established in five large cities (i.e. Seoul, Wonju, Daejeon, Iksan and Busan); and a hydrological one for Flood Control Offices for each of the four major rivers (Han, Geum, Nakdong and Youngsan).

2.3. Involvement of non-state actors

The rise of civil society organisations has been one of the most important political trends in Korea during the last decade (Bertelsmann Stiftung, 2015). Today, environmental NGOs increasingly influence policies, for instance by submitting policy proposals to the National Assembly and the Blue House; by sending written documents and publications to governmental agencies; and by drawing public attention on harmful policies (Kim, 2013).

Similar trends, although seemingly moving at a slower pace, have taken place in the water sector. Large NGOs such as the Korean Federation for Environmental Movement have built up considerable competences in the field of environmental policies, while smaller NGOs have focused on water service provision (Bertelsmann Stiftung, 2015). Consumer associations (e.g. Korea Consumer Agency, Green Consumer Agency) are also important actors with a voice in consultative platforms such as the Water Tariff Committee, River Water Adjustment Councils as well as local price committees. Academia also plays a key role to produce and share technical and scientific information and evidence to build a sound knowledge-base in support of the formulation of policies and practices.

However, the government has a short track record in engaging stakeholders effectively in water-related initiatives. Korean authorities recognise that the effectiveness of water policies is closely related to the involvement and buy-in of water users and citizens. Criticised infrastructure projects, such as the Four River Restoration project, have encouraged public authorities to look for approaches that focus more on public engagement and customer relations. For instance, such changes can be observed in K-water's activities (e.g. the "Tap Water Customer" Councils and the "Water Loves Sharing" initiative) and strategies ("Transparency and Fairness Based in Improved Stakeholders' Trust" is the fourth pillar of K-water 2015 Sustainability Report). However, these efforts remain at an early stage (e.g. K-water's Shared Growth Cooperation Committee was created only in 2015 and its effectiveness has yet to be assessed). Non-state actors remain, more often than not, at the margin of decision making processes on economic instruments.

Several factors come into play to explain the lack of stakeholder engagement. First, there is overall an awareness gap among the Korean population about water financing, which is mostly due to lack of information. The country has made important efforts to open up public data, most recently with its "Government 3.0" Initiative (Box 2.4). Regarding water management, overall, data and information on water quality and quantity tend to be readily available. Most ministries have water-related information on open access on their websites, and produce yearly reports disclosing key statistical data (e.g. MoLIT's annual groundwater report and ME's statistical yearbooks on waterworks and sewerage). In addition, the Water Management Information System (WAMIS) is a common web portal providing information on river basins, dams and groundwater characteristics, and water use. SWM offers real time data on water use and quality.

However, progress is more limited as regards information on water-related economics and finance. Consumers have no information about the source and actual costs of water supply. Above all, water bills do not report detailed information on supply, treatment and distribution costs. As a result, water users cannot understand "who pays for what". Households in particular only have limited knowledge about economic instruments related to water quantity management, which partly reflects trust in the government's capacity to deliver. The lack of knowledge, however, hinders awareness of water-related challenges (risk of shortage, floods or pollution; investment needs) and potential willingness to pay for existing services and future developments. It also raises questions about whether economic tools can improve efficiency of water use.

Box 2.4. **The "Government 3.0" Initiative**

Growth in Korea has been characterised by ICT, with global telecommunication companies such as Samsung and LG, and has been accompanied by government driven open government data policies and e-government strategies, including the recently launched "Government 3.0" initiative.

The Government 3.0 Initiative aims to open up public data and encourage its use by citizens and businesses as well as inside the public administration. In an effort to remove silos among organisations and to develop an inter-ministerial collaborative model, the initiative defines four core values for government institutions, articulated around opening, sharing, communicating and co-ordinating, and puts in place a Public Information Sharing System and a National Future Strategy Centre.

Source: OECD (2015), *Government at a Glance 2015*, Korea Country Fact Sheet; Ministry of the Interior (2014), *Government 3.0*, www.moi.go.kr/eng/sub/a03/Government30/screen.do.

3. Moving ahead

Addressing the institutional challenges identified in Chapter 2 cannot be the responsibility of a single ministry. It requires a whole-of-government approach to take co-ordinated actions at central, basin and local level. As such, further assessment to address the aforementioned institutional challenges would justify a more systematic and in-depth analysis that would bring together all ministries involved in water management – particularly MoLIT, ME and MAFRA – in a concerted approach.

Korea's water policy framework will need to ensure coherence across policy instruments (including with those used for water quality management), as well as policy choices made in consumptive sectors. Positive results could arise from more structural inter-ministerial co-operation. The Water Management Committee performs this role now; however, in the future, it would benefit from operational improvements and enhanced engagement of central government and beyond. Annex 2.A1 provides examples of inter-ministerial bodies that have been created in OECD countries to foster dialogue and actions between public actors in charge of water policy at the central government level. Box 2.5 illustrates how innovative responses to flood risk triggered new institutional arrangements in the Netherlands.

Box 2.5. Innovation and public participation in the "Room for the River" project

Over the centuries, Dutch rivers have had to relinquish much of their space. They are now squeezed between dikes, which in recent decades have become ever higher. In the 1990s, new infrastructure projects advanced slowly, partly due to opposition by inhabitants and "greens", leading to lengthy procedures in court. The imminent threat of flooding in 1993 and 1995 showed that the risk remains: 250 000 people and 1 million cattle had to be evacuated.

An innovative response to flood risks

The government responded to the flooding threat with actions to ensure that flood protection reach statutory levels by 2015 and to improve spatial quality in the riverine area. The objective of the measures was to achieve the required levels of water security while preventing a further increase in water flows in artificialised rivers. This was a shift from the previous dike reinforcement policy that mandated dike improvements where other measures were either inappropriate or unaffordable.

The "Room for the River" programme was completed by 2016. The goal was to give rivers more room to manage high water levels. Measures were taken at more than 30 locations to give the river space to flood safely. Moreover, the measures were designed in such a way that they improved the quality of the immediate surroundings.

Innovative governance

The Room for the River project was carried out in close co-operation between national and regional authorities in both planning and implementation stages of the project. This was essential to ensure completion of the programme on schedule and within budget.

At national level, the Ministry of Infrastructure and Environment and the Ministry of Economic Affairs worked closely together, the State Secretary for Infrastructure and Environment having ultimate responsibility. At regional level, provinces, regional water authorities and municipalities participated in the project. The local government bodies involved residents, business and other stakeholders, including environmental NGOs, in planning and implementation of the project.

The national authorities set central frameworks for safety objectives, spatial planning quality, deadlines and financing. The regional authorities carried out the project within these frameworks. Local governments were better positioned to mesh the plan with regional developments. The Room for the River programme Directorate, under the Ministry of Infrastructure and Environment, bridged national and regional levels. It also monitored the budget while checking, steering and enabling the process and fostering the exchange of information and experience between projects.

The Room for the River administration accelerated the procedures for inquiry, zoning and permitting, acting as a "one-stop shop" for all permits. This reduced the number of (possible) appeal in court considerably.

Local parties wishing to introduce alternative measures could opt for a replacement decision. That option contributed to the sense of equality and was highly appreciated by local parties. It promoted more innovation.

Source: Jan Leentvaar, personal communication.

Stakeholder engagement and policy coherence should also be promoted at river basin level, to overcome unhelpful fragmentation, and to foster basin features in water policies. At the moment, basin organisations reflect national policy making. They are not equipped to foster a comprehensive vision of water issues at basin level and reflect that in national policies. Again, any initiative at that level would require co-ordinated action of several administrations and goes beyond the scope of this policy dialogue.

Korean authorities could also improve the effectiveness of stakeholder engagement. Korea is gradually learning to collaborate and co-operate with civil society organisations and citizens to strengthen the legitimacy of water policy choices. More can be done to upgrade multi-stakeholder platforms currently in place, to emphasise a more practical approach with regular meetings, clear mandates and level of ambition, good information-sharing flows, clear accountability lines and feedback processes.

Annex 2.A1

Inter-ministerial bodies and agencies on water: Examples in OECD countries

In **France**, an "Inter-ministerial mission on water" was set up under the leadership of the Ministry of Ecology and Sustainable Development and more specifically the Water and Biodiversity Department. This administrative commission was created in 1968 in response to the water law of 1964, and brings together all ministries concerned by water policies under the authority of the prime minister. It is responsible for advising the government on any legislative project related to water resources. Its prerogatives on inter-ministerial co-ordination, water management and administration were defined in 1987.

In 2009, **Chile** set up an inter-ministerial committee on water policies to co-ordinate actions between departments and agencies involved in national water strategy. It also advises on strategic planning of water policy in the long term, makes proposals for institutional mechanisms, incentives and guarantees towards the implementation of water policies in rural and urban areas, and adopts the necessary agreements for the implementation of the national integrated water strategy. The committee is led and co-ordinated by the Ministry of Public Works, and has representatives from the General Secretary of the Presidency, the Ministry of Economy, the Ministry of Agriculture, the Ministry of Mining and the National Energy Commission, as well as the National Environment Commission.

In **Israel**, the Water Authority Council created in 2007 is responsible for all decision-making and policy setting by the Israeli Water Authority. It co-ordinates the actions of ministries of Environmental Protection, Health, Finance, Foreign Affairs, and Infrastructure, which used to be collectively responsible for the decision-making over water and sewage. The Water Authority Council was established to alleviate frequent deadlocks that resulted from diverging interests of each agency/ministry. All policies and plans that the Israeli Water Authority or any other Ministry proposes must be presented to the Water Authority Council Forum for approval before they can be passed. The efficiency of the Water Authority Council is contingent upon two criteria – creating equal representation of all interested groups, and ensuring that effective and timely decision-making is their priority. This unifies the responsibility for decision-making on national water and wastewater management and has substantially improved the efficiency and timing of decision-making.

In **Mexico**, the role of CONAGUA, the National Water Commission, is to manage and preserve national waters and their inherent goods in order to achieve sustainable use, with joint responsibility of the three tiers of government (federal, state, and municipal). This decentralised agency of the Ministry of Environment and Natural Resources (SEMARNAT) is the highest institution for water resource management in Mexico, including water policy, water rights, planning, irrigation and drainage development, water supply and sanitation, and emergency and disaster management (with an emphasis on flooding). The Technical Council of CONAGUA is an inter-ministerial body in charge of approving and evaluating

CONAGUA programmes, projects, budget and operations, as well as co-ordinating water policies across departments and public administration agencies. It is composed of the most senior representatives from SEMARNAT, the Ministry for Social Development, the Ministry of Agriculture, Rural Development, Fisheries and Food, the Ministry of Treasury and Public Credit, the Ministry of Energy, the Ministry of Public Administration (SFP), the National Commission of Forestry and the Mexican Institute of Water Technologies.

Annex 2.A1 compiles information from the following sources: Gouvernement français (1987); OECD (2011, 2013), and personal communication with the peer reviewer from Israel.

Notes

1. These include multi-purpose dams, estuary weirs, multi-purpose watercourses, inland transportation for ship, and canals. Dams only for supply of agricultural water not included (see Article 9 – Section 1 – Paragraph A).

2. At the time of drafting the report, the Korean government is considering reshuffling water resources and quality management department, consolidating fragmented water-related departments into one.

3. River Water Adjustment Councils were introduced by amendment of the River Act in 2009. In particular, Article 53 of the Act stipulates that MoLIT may set-up River Basin Adjustment Councils, to be hosted by the Flood Control Offices, which would be tasked with reflecting local residents' views in matters related to river water use and allocation.

References

Asian Development Bank (2014), *A comparative infrastructure development assessment of the Republic of Korea and the Kingdom of Thailand, Mandaluyong City, Philippines*, Asian Development Bank.

Bertelsmann Stiftung (2015), *Sustainable Governance Indicators 2015 – South Korea Report*, coordinated by Thomas Kalinowski, Sang-young Rhyu, and Aurel Croissant.

Dutch water Authorities (2015), *Water Governance: the Dutch Water Authorities Model*, The Hague.

Gouvernement français (1987), Décret n°87-154 du 27 février 1987 relatif à la coordination interministérielle et à l'organisation de l'administration dans le domaine de l'eau (https://www.legifrance.gouv.fr/affichTexte.do?cidTexte=JORFTEXT000000867394&dateTexte=20050530).

Kim, H-R (2013), State-centric to Contested Social Governance in Korea: Shifting Power, *Routledge Studies on Modern Korea*, Routledge, New York.

Kim, K.U., J.K. Koh, et al. (2007), *The dynamics of water quality policy change in South Korea*, CAIWA 2007, International Conference on Adaptive and Integrated Water Management, Basel, Switzerland, NeWater, Institute of Environmental Systems Research, University of Osnabruck.

Ministry of the Interior (2014), *Government 3.0*, Korea.

OECD (2017), *OECD Environmental Performance Reviews: Korea*, OECD Publishing, Paris. http://dx.doi.org/10.1787/9789264268265-en.

OECD (2016), *Water Governance in Cities*, OECD Studies on Water, OECD Publishing, Paris. http://dx.doi.org/10.1787/9789264251090-en.

OECD (2015a), *Stakeholder Engagement for Inclusive Water Governance*, OECD Studies on Water, OECD Publishing, Paris. http://dx.doi.org/10.1787/9789264231122-en.

OECD (2015b), *OECD Principles on Water Governance*, OECD, Paris, available at: www.oecd.org/cfe/regional-policy/OECD-Principles-on-Water-Governance-brochure.pdf.

OECD (2014), *Water Governance in the Netherlands: Fit for the Future?*, OECD Publishing, Paris. http://dx.doi.org/10.1787/22245081.

OECD (2013), *Making Water Reform Happen in Mexico*, OECD Publishing, Paris.

OECD (2012), *OECD Urban Policy Reviews: Korea*, OECD Publishing, Paris. http://dx.doi.org/10.1787/9789264174153-en.

OECD (2011), *Water Governance in OECD Countries: A Multi-level Approach*, OECD Publishing, Paris.

Chapter 3

Economic instruments for water quantity management under the responsibility of MoLIT

This chapter reviews three economic instruments for water quantity management in Korea from a financial and economic perspective, which are under the responsibility of MoLIT. Specifically, it seeks to answer two questions: i) Do the instruments generate the revenues needed to finance the operation, maintenance and renewal of existing infrastructures to supply dam and river water? ii) Can they promote water use efficiency, stimulate private sector investment in water savings, or allocate water where it is most needed? The chapter reviews international experience with similar instruments and suggests options to transition towards economic instruments that actively contribute to water policy objectives.

This chapter does not cover other economic instruments used to manage water in Korea. In particular, the following instruments are out of the scope of this report: water supply tariffs in local authorities, sewage tariffs, water use charge, water quality improvement charge, emission charges, charge for excessive pollution load, environment improvement charge, local resource facility tax, deep sea water use charge and marine environment improvement charge. While the report may occasionally allude to the interplay between the three economic instruments under review and other pricing mechanisms, it does not analyse the other pricing mechanisms listed above.

1. State of play: economic instruments for water quantity management in Korea

This section reviews three water charges administered by MoLIT. They are analysed on the basis of i) their capacity to generate the revenues needed to build, renew, operate and maintain water storage and distribution systems; ii) their capacity to drive water users' behaviour and water demand; iii) and the institutional arrangements that govern their administration.

The **river water use fee** is levied on water users (energy generators, industry, and domestic users [1]) who abstract water from a river. It is levied by provinces and metropolitan cities (including Seoul) and earmarked for river maintenance and repair expenses. The rate is set by the River Act Enforcement Decree nationwide. Local authorities are expected to pay the river water use fee, even when they are endowed with vested amounts of water; however, they themselves collect the revenues and do not report explicitly about revenues collected and expenditures covered.

Two points deserve attention:

- In principle, permission to use river water is delivered by flood control offices in each river basin (under the responsibility of MoLIT), but the fee is levied according to local authorities' ordinance. Some local authorities calculate the fee on the basis of the entitlement of river water while others refer to actual water use. In 2015, the Civil Rights and Anticorruption Committee recommended that the basis on which the fee applies be clarified.
- Revenues generated by the river water use fee are collected by provinces and metropolitan cities (including Seoul). Anecdotal evidence from interviews during the fact-finding mission suggests that revenues cover roughly half of the expenditures related to river management. Collection rates are low (the fees are collected by entities that are the main water users): river water use fee can be waived for non-profit public service, as defined by Presidential Decree. Local authorities report the revenues generated by the river water use fee to MoLIT, but full details are not reported.

The **dam water tariff** is a uniform volumetric tariff levied on water users who abstract water from a dam via a contract with K-water. The tariff is set nationwide with a view to recover the cost of the construction and operation of dams used to store and supply water. Revenues are collected by K-water and earmarked for dam management. According to the River Act Enforcement Decree, the dam water tariff and river water use fee for industries and domestic uses are the same.

The **multi-regional water tariff** is a two-part tariff set nationwide to recover the costs of supplying water through multiregional systems. It covers the cost of abstracting water (river water use fee or dam water tariff) and the cost of treating and supplying water to the user. The fixed part of the tariff should not exceed 30% of the total. Different tariffs apply to different water quality grades (raw water, settled water, purified water). Revenues that correspond to abstraction are transferred to the relevant authority (provinces and metropolitan cities in the case of river water use; or K-water in the case of dam water). Revenues generated in addition to water abstraction costs are earmarked for the operation of multi-regional water supply systems. They are supposed to cover the O&M costs of the service and the costs of renewal or upgrade.

Special tariffs (reduction of or exemption from a fee) apply to maintain environmental improvement flows or encourage river bank filtration (dam water tariff), when dam or multi-regional water is below certain quality standards, or for customers with long-term contracts (2-5 years), those who install water reuse systems, or socially under-privileged groups (multi-regional water tariff).

At local level, local authorities set water supply and wastewater tariffs, which vary across the country. Tariffs partially reflect the charges paid by local authorities to access water either directly from a river, a dam, or from procuring multi-regional water services (approximately 20% of the water bill).

According to the River Act, water fees and charges are waived when water is used for environmental improvement, or fire and disaster management. Irrigators are exempted from river water use fee and dam water tariff, based on a Presidential decree.[2]

Table 3.1. **Charges on water abstraction in Korea**

	Who collects	Who pays	Who is exempted
River water use fee	Provinces and metropolitan cities	Municipalities, industries	Irrigators, etc.
Dam water tariff	K-water	Municipalities, industries	Irrigators (they seldom use K-water reservoirs), etc.
Multi-regional water tariff	K-water	Municipalities, industries	

1.1. Current expenditure and financial flows

Tariffs for dam water and multi-regional water systems are set by MoLIT in consultation with the Minister of Strategy and Finance.[3] The water tariffs (at the time of the drafting the report) are shown in Table 3.2. For reference, the average tariff for water supply for domestic use is 486.7 KWR/m³ as of 2014; it varies across regions.

Table 3.2. **Water charges for water quantity management in Korea**

2016 – KRW/m³

River water use fee	Dam water tariff	Multiregional water tariff for raw water	Multiregional water tariff for settled water	Multiregional water tariff for Purified water
52.7	52.7	233.7	328.0	432.8

The current cost recovery rate is low (see Table 3.3): tariffs for dam and multi-regional water supply (and for water supply and sanitation services – not covered by this policy dialogue) hardly cover operation and maintenance (O&M) costs and partial capital charges. This indicates that dam or multi-regional water is supplied at a tariff level that is cheaper than the full supply cost.

Table 3.3. **Cost recovery ratio for dam water and multi-regional water supply**

					2016	
	2012	2013	2014	2015	Before tariff increase	After tariff increase
Dam water	79.3%	91.2%	88.1%	88.2%	82.7%	86.7%
Multi-regional water supply	82.6%	84.5%	84.8%	83.8%	84.3%	88.3%

The tariffs reviewed in this section are not only low, they are remarkably stable over time. Looking into the change in water tariff specifically, average tariffs for water supply and sanitation have increased by 17.3% and 63.2% respectively between 2005 and 2013; over the same period, the rates of water use charge have increased by 14.3% to 41.7% depending on the water sources; and the dam and multi-regional tariffs (raw water) have only increased by 4.9% and 2.7%, respectively.[4] The tariffs have been adjusted 3 times in the last 10 years: in 2007, multi-regional water tariff decreased by 0.4%; the price of settled water decreased by 5%. Both tariffs increased by 4.9% in 2013, and by 4.8% in 2016. At the same time, Consumer price index increased by 27.5%. Change in dam and multi-regional water tariffs has been restrained by the government's price stabilisation policy (which applies to public utility charges) (MoLIT and K-water, 2016).

One reason to keep tariffs low in Korea is to control inflation and to address social concerns. This reasoning is partially valid at best. First, 1% increase in dam and multi-regional water tariffs affects Korean consumer price index by 0.0013%. Second, while the

Figure 3.1. **Water tariff. Trends over 2005-15**

KRW per m³

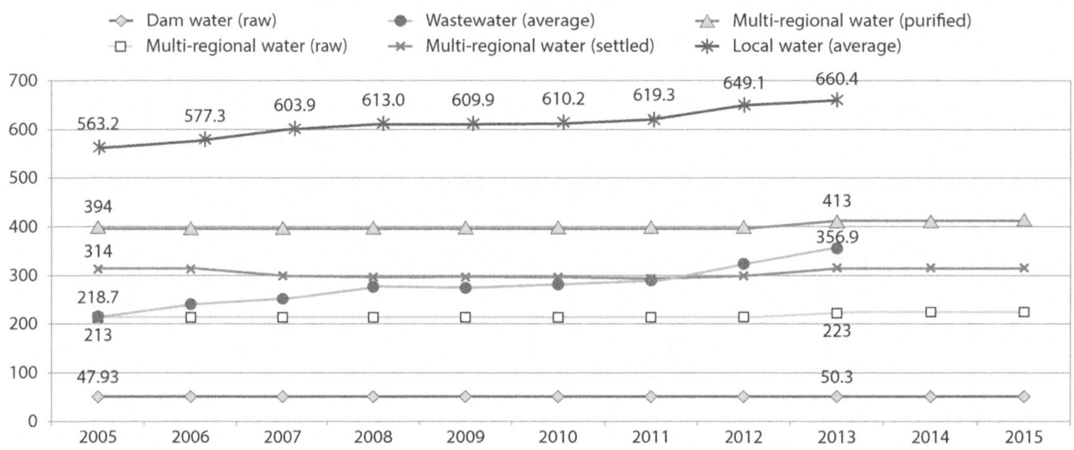

Sources: MoLIT and K-water.

Figure 3.2. **Water use charge. Trends over 2005-15**

KRW per m³

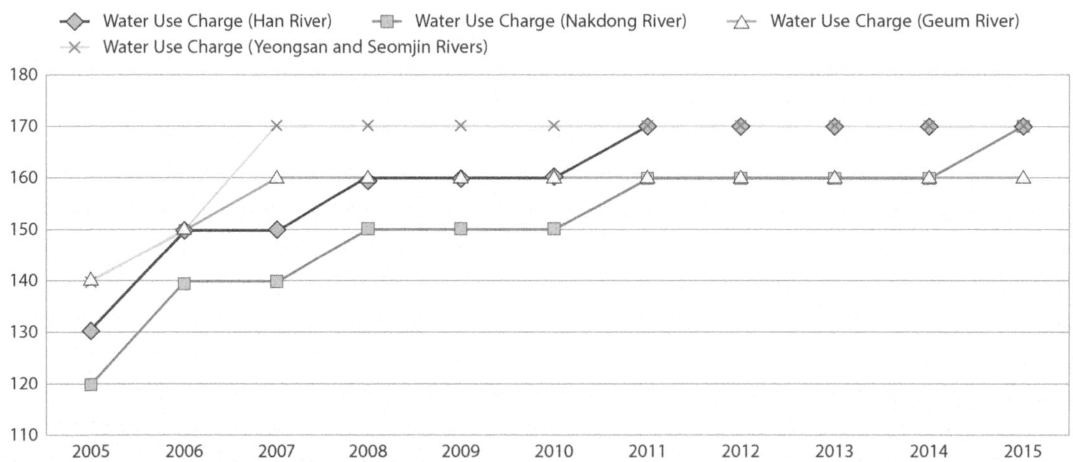

Sources: MoLIT and K-water.

tariffs under study only marginally affect final users' bills, water bills in Korea as a share of disposable outcomes of the lowest decile of the population are the cheapest in OECD countries (OECD, 2010). This suggests that 90% of water users could afford more expensive water tariffs. Preliminary discussions with consumer groups in Korea suggest that the willingness-to-pay may be higher than assumed by local authorities, as long as water taste and safety and service quality improve. Water bills represent only a fraction of households' expenditure on water supply and sanitation: 48.7% of households use a water purifier. The domestic market for water purifiers was estimated to be KRW 2 trillion 2014 (Nielsen Korea).

In such a context, the long-term reproduction of physical assets is not secured, as only a part of the cost is recovered through revenues from the instruments under review. Financial sustainability will depend on the capacity of water authorities to secure additional funding from public budgets, essentially at central level. Preliminary discussions with MoSF indicate that this may not be easy. This situation contributes to postponing the replacement of capital stock, resulting in aging infrastructures and unnecessary wastage of water and expenses. As shown in Table 3.4, aging infrastructures in multi-regional supply systems over 20 years are expected to reach 49.6% by 2020 and 79.6% by 2025 out of total facility capacity.

Table 3.4. **The proportion of aging water infrastructure over 20 years**

Water infrastructure	2014	2020	2025
Multi-regional water supply system	26.3%	49.6%	79.6%
Local water supply system	30.6%	37.7%	45.9%

Source: MoLIT and K-water.

The three tariffs reviewed in this section are therefore unable to generate appropriate levels of funding and to alleviate the government's financial burden. They put financing of dam and multiregional water at risk, since competition intensifies among a number of demands on public budget such as social welfare, equality improvement, education and economy stimulus.

1.3. Shortcomings as economic policy instruments

The three charges reviewed in this section do not contribute to economic efficiency. The rates are too low to affect the behaviour of final users. In addition, the water bill that domestic users receive does not directly reflect the charges,[5] and local authorities may decide to defer the impacts of increases in water charges on the tariff to end users. As a result, the charges do not effectively discourage wasteful freshwater use.

Moreover, the rates are uniform across the country, which does not allow them to reflect local conditions, such as flooding or scarcity, which are under MoLIT's authority. This prevents policies from discouraging water use in water scarce areas and encouraging water intensive activities where water is abundant.

Finally, the design of the three instruments prevents them from allocating water where it creates most value. The instruments are well-adapted to a policy that results in supply augmentation financed through (almost unlimited) public budgets. They cannot support a policy that promotes water use efficiency, minimises investment in new infrastructure, and effectively allocates scarce public finance.

The three charges under review can contribute to water security in Korea if they finance well-designed investment and expenditure programmes. As mentioned above, revenues from the river water use fee are collected by local authorities and earmarked to cover river maintenance costs. In practice, however, local authorities do not report explicitly on the expenditures financed by the river water use fee. As a result, it is practically impossible for MoLIT to trace how these revenues are spent, to critically assess local authorities' spending choices, to suggest alternative spending options that could create more value, or to hold local decision makers accountable for misuse. Anecdotal evidence suggests that citizens do not experience immediate benefit from these expenditures.

Revenues from dam and multi-regional water tariffs only contribute to water security as long as K-water expenditure programmes are cost-efficient. Such an investigation requires an independent economic and financial review, which is beyond the scope of this report.

1.4. Governance shortcomings undermining the use of economic instruments

Chapter 2 highlighted three governance challenges that can affect the operation of the three water charges under review: a lack of horizontal co-ordination at national level, a vertical co-ordination gap between the central government and subnational authorities, and limited opportunities for stakeholder engagement.

Horizontal co-ordination at national level

The multitude of institutions involved in water policy making and management raises the complexity and costs (e.g. administration costs) of designing and implementing water policy instruments, including water charges. For instance, the three pricing instruments under review are set to recover costs of bulk water supply services or rivers/dams maintenance/operation, with no consideration for water resources management and water allocation. This institutional organisation affects policy coherence. Water quantity management spills over and is affected by choices made in other policy fields (i.e. agriculture, energy, urban development, public finances), which are often responsible for growing pressures on water resources. The multiplicity of actors involved in water management at central level, and the lack of institutional incentives in Korea (e.g. common objectives, performance indicators, etc.) for more systematic horizontal co-ordination between different policy fields, may make it difficult to exploit synergies or encourage cross-cutting agendas (for instance, setting charges that signal the opportunity cost of augmenting water supply by building new dams, that reflect water scarcity or contribute to allocating water where it creates most value).

Vertical co-ordination

In Korea, charges for water quantity management are set at central level. On the one hand, this approach is well-suited to provide access to bulk water at the same price across the nation. On the other hand, this approach cannot reflect local conditions and leaves communities in different locations to face a range of challenges (such as limits of water availability, difficulties accessing the resource, level of equipment or infrastructure, level of economic development, etc.). To be effective, water charges need to be responsive to these differences. Subnational authorities are best placed to understand local needs and to engage relevant interlocutors. River basin organisations under MoLIT are in the right position to help the government reflect territorial specificities in spending choices, identify necessary projects, and develop investment plans at basin and sub-basin levels.

However, river basin organisations under the responsibility of MoLIT are not represented in the Water Tariff Committee. In practice, basin-specific issues related to economic instruments (which may relate to different situations in terms of water scarcity, or impacts on tariffs on regional economies) can only be reported through MoLIT. They may eventually be mentioned by consumer groups in local governments. They only rarely are reported to the Committee (see Box 3.1).

Box 3.1. **Engaging stakeholders at basin level: The case of France**

Stakeholder engagement in France has played a critical role in establishing river basin committees. Each of the seven French basin committees is chaired by an elected local official and made up of representatives from local authorities (40% of seats); users and associations including industries, regional developers, farmers, fishermen, tourism and nautical activities, electricity producers and water suppliers (40%); and the central government (20%).

As members of the basin committee, these stakeholders orientate the water policy priorities in their respective river basin. They prepare the master plan for water development and management (*Schéma directeur d'aménagement et de gestion des eaux*), which is approved by the central government and monitor implementation.

Stakeholders also formulate the priorities for each river basin institution regarding tax levies, in particular abstraction and pollution charges, within the limits determined by the national water law. The basin committee votes on the multi-year action programme, which sets the priorities and methods for financial assistance to fund the implementation of the river basin master plan.

Sources: OECD (2015a), *Stakeholder Engagement for Inclusive Water Governance*; Official website of the French Water Agencies, www.lesagencesdeleau.fr.

Stakeholder engagement

Three ministries are involved in the procedures for setting dam, multi-regional and river water charges: MoLIT, MoSF and MoI (which directs local waterworks management). Together with K-water, MoLIT and MoI deliberate on tariff changes within the Water Tariff Committee. This consultative committee also includes 11 consumer representatives from the domestic and industrial side, academics and accounting experts (two of them are local authorities). These stakeholders are appointed based on recommendations from civil society organisations, Korean Industrial Complex Corporation (an affiliate of MoTIE), and the MoI. They are then invited to participate by MoLIT.

Stakeholders interviewed in the framework of this project stressed that current procedures could be improved regarding the selection of members and operation of meetings in order to reinforce representativeness, legitimacy of the deliberations and accountability of those who take part.[6]

2. Towards economic policy instruments for water use efficiency

The section explores how the economic function of pricing instruments to manage water quantity in Korea could be strengthened. It initially focuses on the three instruments under review, in particular with a discussion on how the instruments can better contribute to financing objectives It then explains the rationale for abstraction charges that signal water scarcity and the opportunity cost of using water. It discusses implementation issues and shows how they have been addressed in a range of countries. The following section will discuss how international experience can inspire policy reforms in Korea.

2.1. Making the best of financing potential of economic instruments under MoLIT's responsibility

The analyses in the previous section indicate that the performance of the *river water use fee* as a financing instrument for river maintenance is weak. Enhancing the performance of this instrument requires careful consideration of both the revenue raising capacity and expenditure side. From a revenue raising capacity, a basic requirement would be that revenues collected from the fee are monitored. This is a requisite to appreciate compliance (those who use river water pay the fee), target information campaigns, as well as control and possibly sanction penalties. From an expenditure perspective, as the revenues collected are earmarked for expenditures at local level, MoLIT should have the ability to monitor and assess the relevance and effectiveness of the expenses financed from the river water use fee. This would allow MoLIT to support local authorities in planning and selecting expenditures that best contribute to river maintenance and help local authorities build implementation capacity.

When these requisites are in place, a revision of the river water use fee could be considered, based on an explicit programme of measures that contributes to river maintenance. MoLIT's regional branches could review such programmes, assess their opportunity and their efficiency, and ensure they contribute to river maintenance and national policy objectives at least cost. There are benefits to involving citizens in the design of expenditure programmes: when citizens see that such programmes actually improve their water security or quality of life, they are more willing to cover the costs.

Similarly, the analyses in the previous section show that the revenues from the *dam water tariff* and the *multi-regional water tariff* fail to cover the costs of operating, maintaining and renewing the existing infrastructures. Both tariffs should cover the costs of supplying the service (management of dams and multi-regional water supply facilities). This can be difficult, in particular as regards multi-regional water tariffs.

The cost of bulk water supply varies considerably depending upon the location of the delivery point within the bulk water supply infrastructure system. From an efficiency point of view, it is therefore not appropriate to apply a uniform bulk water tariff to all water supply service providers. Tariffs should be set to reflect the costs of supply at the abstraction point in the bulk service provider's system. Reflecting costs is central to securing sustainable finance for service providers. From an equity point of view, Korea's policy has been to set tariffs nationally so that every user has access to water under the same condition. How much do national tariffs contribute to an equitable access to water will be discussed in the next section.

Ex-ante price regulation of specific access service is desirable. The price paid to the bulk water supplier is usually reviewed through a public process including the following steps: i) analyse bulk water supply to set out the expected costs to deliver agreed-upon capital work programmes, service standards and forecast volumes for each prescribed service; ii) review capital expenditure, operating expenditure and forecast demand; and iii) consult stakeholders. Box 3.2 illustrates how this regulation is organised in the UK.

Box 3.2. Determination of relevant cost for bulk water: The UK example

Identifying relevant cost is a controversial stage of bulk water pricing. The main principle used by Ofwat – the economic regulator of the water sector in England and Wales – for identifying the cost of a bulk supply is to consider that the costs incurred should relate to the services being provided. Ofwat first identifies the relevant set of services that are provided under the bulk supply agreement and then assesses the cost of these services.

How are the costs reasonably associated with the provision of the relevant services assessed by Ofwat? In line with the provisions of sections 40 and 40A of the Water Industry Act 1991 (WIA91), both already incurred costs, and costs likely to be incurred in the future, can be considered as relevant. For example, already incurred costs can be particularly relevant if a bulk supply agreement is being changed or terminated. This is because investment decisions, potentially involving significant cost commitments, may have been taken based on the existing agreement.

In determining costs, a key question is whether (and to what extent) bulk supply determinations should use: the average cost of supplying other customers ("average cost"); or an assessment of the costs associated with a specific customer ("specific cost"). If a service is provided using a network, it may be difficult to distinguish between the costs associated with providing an individual customer with a service, and the costs of other services. When the provision of services involves the same network there may be "joint" costs, "common" costs, or network benefits.

In such cases, Ofwat typically considers *practicability* and *proportionality* when determining the extent to which bulk water pricing should try to reflect the costs of serving particular customer requirements.

Here are a few examples provided by Ofwat.

- If a network has a large number of relatively small users, a common charging scheme that involves a significant degree of averaging can provide a practical and proportionate means of reflecting the costs of provision. This is despite the fact that there may be some relevant differences in circumstances between particular customers that will not be closely reflected in the resulting charges.
- If there are fewer users, and the network interactions are more limited, the balance of arguments may be more in favour of a specific cost assessment.

The *practicability* and *proportionality* of different approaches to the assessment of cost will partly depend on the availability of information. When considering issues of practicability, it is important to recognise that the availability of information is likely to evolve over time. This means that Ofwat retains the possibility of changing the way costs are assessed depending upon available information.

Source: Adapted from Ofwat (2011), *Bulk supply pricing – a statement of our policy principles.*

A basic principle is that those who generate costs or who benefit from access to reliable water sources should cover the costs. Failing to do so raises equity issues, as future generations will have to foot the bill, either through decayed infrastructure or increased public debt. Typically, all things being equal, industry or municipalities should have an incentive to develop in regions where water is abundant and competition between water users is minimal; and they should cover the costs of a reliable access to water. In the Netherlands, urban development only takes place after exposure to water risks (essentially flooding) has been assessed. The Water Assessments show how an urban development affects water security; they substantiate discussions at community level, on the opportunity of the development and the allocation of costs for mitigating water-related risks for that urban development.

Box 3.3. **Reflecting water issues in land development in the Netherlands**

Spatial planning decisions are significant for water management, as decisions on the location of industry and urban development affect the costs of ensuring water security (supplying water and protecting against floods and risks, essentially). In the Netherlands, the Water Assessment (In Dutch, *Watertoets*) serves as an instrument to make transparent the impacts on water management of spatial development and to help to discourage urban development in unfavourable locations. The Water Assessment can best be understood as a communicative process in which water managers advise on the consequences of land use developments for the water systems, and vice versa.

The Water Assessment plays an important role in the co-ordination of water plans on the one hand, and municipal structure plans and land use plans on the other. While the instrument was initially designed for flood risk assessment, it can be adapted to reflect risks of scarcity or the costs of supplying water in a water scarce region.

Source: OECD (2014b), *Water Resources Governance in the Netherlands*. http://dx.doi.org/10.1787/9789264102637-en.

Just as dam and multi-regional water tariffs should cover the costs of supplying water, they should include an incentive for the operator to manage the infrastructure in an efficient way. If they do not, there is a risk that the supplier overspends in infrastructure development. This is one of the major roles of the economic regulator in England and Wales (see Box 3.2). In the Netherlands, recent discussions on financing flood protection by the Regional Water Boards indicate that expenditures should not be driven by the capacity to raise funds (OECD, 2014b). The point is not only to check whether an investment is properly managed, it is also to verify that this investment is needed. This is essentially a regulatory issue. Several options can be considered to oversee decisions about building additional infrastructure and the best options to minimise investment and O&M costs while maintaining an adequate level of water security. The trust citizens have in regulatory arrangements is critical for their support and willingness to pay.

Any suggestion to raise the fees and tariffs under the responsibility of MoLIT should be accompanied by i) a realistic assessment of the potential benefit in terms of water security and additional service for final users (industry, households, the environment), and ii) a thorough assessment of the social consequences of such a raise, essentially in terms of affordability. Based on existing information, the impact on affordability are likely to be minimal, if not negligible, as the instruments represent a very small part of the final water bill and of households' revenues. This is an additional reason why this information should be disclosed.

2.2. The rationale for abstraction charges

In addition to raising funds to operate, maintain and renew existing infrastructure, water fees and tariffs under the responsibility of MoLIT can potentially contribute to water quantity management. At present, revenues collected via the instruments under discussion do not reflect the scarcity value of water and therefore do not contribute this additional goal (see the definition of the opportunity cost of using water, below). There could be benefits in transforming existing instruments into abstraction charges that reflect the scarcity value of water, which are addressed in this section along with examples of international good practice. In addition, this section seeks to put forth a roadmap to implement the suggested reform, acknowledging the sensitivity of such a reform and addressing particular issues or concerns that may arise in the Korean context.

Introduction to abstraction charges

In a perfectly competitive market, the price given through interaction between buyers and sellers is efficient, and the optimum allocation of water is automatically achieved. But with water, specifically due to its environmental uses and its associated externalities (water as a common pool resource and a source of pollution), prices have to be set to correct inefficiencies resulting from individual behaviours.

Availability of water depends on location, time and probability or "state of nature". Economic theory tells us that efficiency can be achieved with contingent market and/or pricing provided that the public good provision is fixed with regulation. Externalities related to water abstraction should be taxed using a Pigou rate reflecting the externality's social marginal cost. The Pigou tax should thus differ in time and space, and should be contingent on uncertain features such as weather conditions (temperature, water supply).

Pigou taxes are difficult to implement in practice because of transaction costs or other market failures such as asymmetric information. As a result, when setting water charges, decision makers have to determine a good balance between mimicking contingent pricing and lowering transaction and implementation costs. Decision makers have also to assess the possible trade-offs between incentives and cost recovery: an efficient water tax or charge may undermine its own function as a source of revenues if high water prices reduce water consumption. This last effect highly depends on price sensitiveness of water users (price elasticity).

Economic principles

Figure 3.3 assembles the various costs associated with water abstraction and use. The costs on the figure are indicative only and not at scale. The full supply cost aggregates the costs incurred by the use of infrastructure and appliances to abstract water. The full economic cost adds to that the opportunity costs of using water and economic externalities. The full cost adds to the economic cost the environmental cost of using water.

Abstraction from groundwater and surface water is costly. It requires some infrastructure or equipment to pump and distribute water. This cost is usually set by water users and is sometimes called the use cost or the **full supply cost** (see Figure 3.3). Since it is relatively easy to quantify, this report does not discuss in details how to set full supply cost. Rather, the focus is on two other types of costs that are more difficult to estimate, and may be less obvious to both the general public and the users of the water resource. Figure 3.3 describes a situation where fees and tariffs do not cover the full supply cost of dam or multi-regional water.

The **environmental cost** is the cost of the damage induced by water abstraction on the natural environment. For example, too much groundwater abstraction can cause saline intrusion in coastal aquifers while the ecosystem of rivers and other surface waters may be degraded when river flows become too low.

Figure 3.3. **The composition of the various components of water costs**

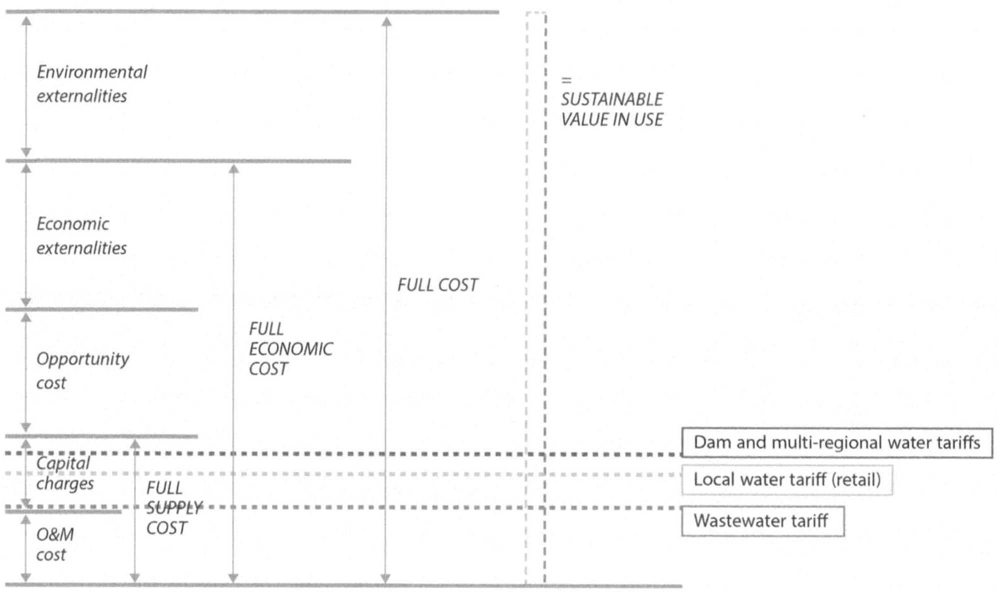

Source: Adapted from Rogers, P., R. Bhatia, A. Huber (1998), *Water as a Social and Economic Good: How to put the principle into practice.*

In regions where water is scarce, water abstraction will also impose **opportunity costs** of using water (or resource, or scarcity costs), which can be seen as foregone opportunities of alternative water uses. These costs are incurred when one water user affects the use of the resource by another user. For example, greater abstraction of water by a city might affect the quantity of water available to downstream irrigators, thus imposing costs on these users. Technically, the opportunity cost is defined as the value of the water in its highest value alternative use.

Figure 3.4 provides an illustration of resources costs (RC) and the optimal allocation of water under scarcity. The horizontal axis shows the available water during a year and the vertical axis the value of water in KRW. Water demand for user 1 and 2 are represented by the demand curve D_1 and D_2.[7] In a normal year – i.e. available water is represented by the full horizontal axis – the demand curve defines the marginal benefit for users 1 and 2; optimal usage is achieved when users 1 and 2 get no additional benefit from additional consumption. Their optimal water consumption is therefore equal to the quantities q_1^0 and q_2^0.

As illustrated in Figure 3.4, during a normal year both users can consume their desired quantities and there is water left in the system (distance between q_1^0 and q_2^0). In a dry year – illustrated by the shrinkage of the available water represented by the red arrow – the demand for water is the same, but there is now competition for the scarce resource. It is therefore necessary for a regulator to come up with an allocation mechanism for the available water. One option is to prioritise user 2 and allow user 2 to consume its desired quantity q_2^1. However, as illustrated by the graph, at that level, the value of water is higher for user 1 than for user 2. It follows that such an allocation regime is costly for society: society does not reap the full benefit of using water where it creates most value. The optimal allocation for society is when the marginal benefit of water is equal among the users; this situation corresponds to $q_{1,2}^*$: the intersection of the demand curves sets the opportunity cost. Hence, in theory, optimal use of water can be achieved if a regulator sets a resources cost at a level equivalent to $q_{1,2}^*$.

Figure 3.4. **Opportunity costs and optimal allocation of water under scarcity**

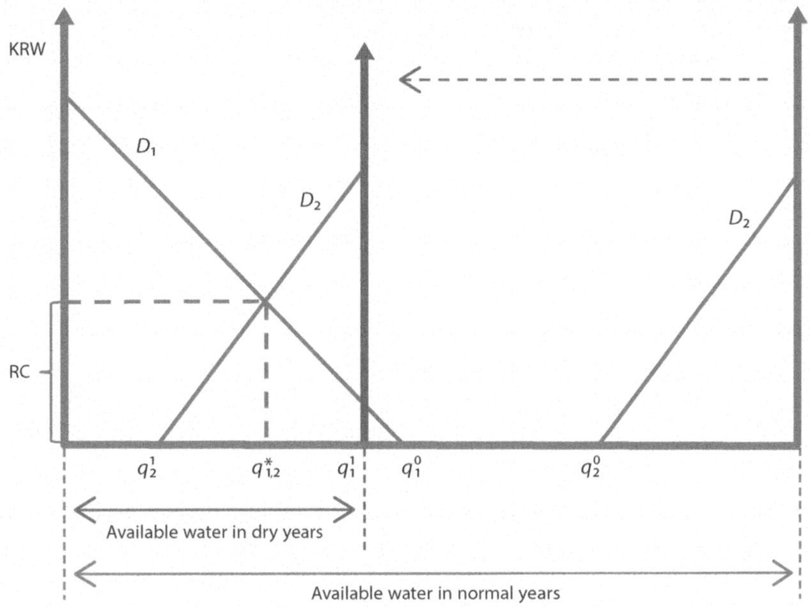

Source: Henrik Andersson, personal communication, for this report.

Opportunity and environmental costs reflect the externalities that water abstraction by one user causes to other users and the environment. From an economic point of view, because these costs are not reflected by market mechanisms, the intervention of authorities is justified. It makes sense for governments to set up charges so that water users are able to internalise these costs when deciding on the volume of water to be abstracted. Charging users for the environmental and opportunity costs of using water is essential to recover the full cost of using water (see Figure 3.3; see also Article 9 of the EC – Water Framework Directive WFD – 2000/60/EC) and achieving sustainable water resources management. Water charges that reflect the environmental and opportunity costs of using water enhances the efficiency of water allocation across users, where "efficiency is concerned with maximising the welfare that is obtained from a resource by allocating it to its most valuable economic use" (OECD, 2011).

Water abstraction charges may however raise equity and competitiveness issues, for example if some of the water users (e.g. low-income households, some specific industries) are made substantially worse-off after the implementation of the charge. The economic principle is that charges should be related to allocative efficiency. Equity concerns can be fixed with targeted social transfers. Abstraction charges can be adjusted to address equity issues when transfers are not possible. In particular, equity prescribes applying the polluter-pays principle which implies that the polluter pays for the damage he or she causes through Pigou taxes (and victims are compensated for the damage it causes).

2.3. Design and implementation issues

The measurement of both the environmental cost and the opportunity cost is a difficult exercise. Assessing the environmental cost implies valuing the cost of the damage caused by the abstraction of water on the ecosystem. Some of these impacts may be visible only in the long run (for example impact on the fish population or the biodiversity) and are difficult to monetise because environmental benefits are not currently priced. Therefore, it requires bio-economic studies to be conducted in order to estimate the full marginal value of environmental flow in each watercourse. Such studies are difficult to undertake since they require expertise from various sciences, are resource-intensive, and their findings may not be easily transferable since estimates are going to be, in most cases, site-specific.

Assessing the opportunity cost requires measuring the highest value of the possible use of water that could not materialise because of the volume already abstracted. For instance, what would have been the revenues of an efficient farmer if the city upstream had not abstracted additional water? Another option is to calculate the marginal cost of supplying the additional volume of water required to meet the needs of the farmer (see the case of Israel in the next section).

Choosing the level of the charge such that it covers the exact cost of the environmental and resource costs is an almost impossible task for public authorities. The best they can do is to approximate these costs by relying on impact studies and by partially accounting for the difference in costs depending on the uses, the location, and the time of the year.

The volume of water abstracted is usually chosen as the basis for charging; however, this approach raises two potential issues. First, part of the water abstracted is returned to the environment. If the water that is returned is of the same quality, and if it is returned where it was pumped, then it is not abstracted water that should be charged but rather water that has been used (and not returned to the environment in the catchment). This, however, may be difficult to quantify since it would require a meter to measure water abstraction and a meter to measure water that is returned to the environment. Second, a meter is required to measure the volume of water that is abstracted. Meters are not always installed, especially in the agricultural sector. If abstracted water cannot be metered, then authorities can rely on other bases for taxation to approximate water use. Examples include charging hydroelectricity production on a per megawatt-hour basis.

The form of the charge matters. Authorities may impose a volumetric charge that remains the same whatever the volume extracted or design charges such that the per unit price varies with the total volume of water extracted (decreasing or increasing block schemes). Based on economic efficiency principles, the latter would be justified if the marginal environmental cost or marginal opportunity cost varied with the volume of abstracted water.

The implementation of the charge may also impose transaction costs on water users and managing authorities. Water users subject to the charge usually have forms to fill to indicate the volume of abstracted water and monitoring (control) by managing authorities does not come at zero cost.

The incentive role of the charge on water abstracted or used will depend on its level and the elasticity of water use to a change in water prices. Domestic users are commonly found to be sensitive to prices but the elasticity of water use to price changes is, in most cases, found to be rather small (in the range -0.1 to -0.4). It follows that the charge needs to be substantial to induce a change in water users' behaviour. In France, for example, the abstraction charge represents only around 1.3% of the water price paid by households and is thus unlikely to make them change their behaviour (French Ministry of Ecology, Sustainable Development, and Energy, 2012).

2.4. Empirical estimation of the externalities and opportunity cost(s)

Because environmental benefits are typically not priced, non-market valuation techniques (e.g. contingent valuation, travel cost method) have to be used to estimate the economic value of environmental benefits of water flows. Assessing the resource cost is also difficult since it requires measuring the benefit induced by water for all types of use (agriculture using irrigation water, drinking water supply, hydropower etc.), with benefits that commonly vary from one season to another (opportunity costs are usually higher when rainfall is low).

Non-market valuation techniques can broadly be defined as either belonging to revealed (RP) or stated-preference (SP) methods. The former refers to methods that use market data to elicit monetary values, either as how much individual would be willing to pay for the good or service of interest, i.e. their willingness to pay (WTP), or how much they would require in compensation for being exposed to a loss of welfare, i.e. their willingness to accept (WTA).[8] The latter refers to methods that use hypothetical scenarios to estimate individuals WTP or WTA. The methods have been used for a large range of non-market goods, in addition to water (e.g. air pollution, noise levels, health risks, etc.) and have improved (see Box 3.4). Examples of RP techniques are the hedonic regression technique and the travel cost method. The former can be used to estimate the value of clean water by examining how property prices depend on proximity to clean water (everything else equal), whereas the travel cost method mainly has been used to estimate recreation values, such as access to kayaking, fishing, etc. Examples of the SP techniques are contingent valuation and discrete choice experiments: based on hypothetical scenarios they are able to elicit individual WTP for a large range of goods and services like access to safe water, better tasting water, secure access to water, and recreational values.

Box 3.4. **Non-market valuation of environmental flows in Australia**

The River Murray and the Coorong are a unique ecosystem which provide habitat for breeding birds, fish, and vegetation. However, decreasing environmental flows during an extensive drought contributed to over-extraction and declining inflows mean that the area and its habitat have been in decline. One method of estimating the value of environmental flows is to design a survey which asks people their willingness to pay for improvements in environmental quality, using this as a measure of the value people put on the environmental services provided.

In order to estimate the value of these environmental flows in the Murray River and the Coorong, MacDonald et al. (2011) designed a survey that was sent out to over 3 000 Australian residents. The survey described the impact of low environmental flows on waterbird breeding habitat, native fish populations, and healthy vegetation in the area, and set out ways of improving environmental quality by purchasing water user's rights from willing sellers, investments in irrigation efficiency, and habitat rehabilitation, together with the costs of these policies. The survey then asked respondents to choose between various policy options that had different environmental impacts and different costs.

Through a statistical analysis of the results from the survey, MacDonald et al. (2011) found that Australian residents were willing to pay substantial amounts to improve the quality of the Murray River and Coorong indicating that the value of environmental flows in the area is significant. Specifically, total willingness to pay (in present value terms) to increase the frequency of waterbird breeding from every 10 years to 4 years, to increase native fish populations from 30% to 50% of original levels, to increase the area of healthy native vegetation from 50 to 70%, and to improve waterbird breeding habitat quality in the Coorong was AUD 13 billion (USD 14 billion). The authors stress that, due to the uniqueness of the Coorong, this value cannot be used to estimate the value of other watercourses in Australia, and further surveys are required.

Source: MacDonald et al. (2011), Valuing a Multistate River: The Case of the River Murray, *The Australian Journal of Agricultural and Resource Economics*, Vol. 55.

Non-market valuation methods have been used in Korea to derive WTP for different water services. For instance, whereas a few studies have estimated households' WTP to secure (reliable) access to water (e.g. Park and Park, 2007; Chae and Kang, 2011), there is a rich literature estimating the WTP for improved water quality (e.g. Kwak, 1994; Kim et al., 2001; Pyo and Choo, 2014). Several studies have used the traveller cost method to elicit WTP related to recreational water activities (e.g. Lee et al., 2015; Pyo, 2015).

The description above mainly covers the environmental externalities, i.e. quality and recreational value of water. The non-market valuation techniques can also be used to monetise opportunity costs of using water. They include either production function approaches or intermediate good (value added) approaches where the production cost and income will be affected by access to water. For instance, assume a government programme that provides water for farmers. The income differences between those having access to the government programme can be compared to those not having access to the programme and this difference would reflect the benefit of the programme.[9] An alternative to estimate the opportunity cost of using water is to rely on markets for trading water rights. Such a market would provide a clear indicator of the value of water, but the WTP would depend on the amount of rights provided by the regulator. For instance, in a situation with a too generous regulator the value of the water rights would be low and would not be a good signal of the value of water in a water scarce context.

The monetary estimates of water and its externalities will be context dependent, i.e. they will depend on where and when, and under what regulatory system they are estimated. Hence, it is likely that values will vary both in time and spatially. Therefore, when added to the user cost, which may be different from one time and place to another, the monetised values for the environmental externalities and the opportunity cost may either increase or decrease the variation in values.

2.5. A review of international experiences

Water abstraction charges are widely used in OECD countries. They can serve water policy objectives under a range of circumstances. This section captures some lessons from the experience of several countries. The cases of France and the UK (England and Wales) are covered in more details in appendices to this chapter.

The French and English and Welsh cases illustrate i) how charging for water abstraction relates to zoning of water resources based on resource availability; ii) the role of stakeholders engagement, as the charge does not merely result from a technical calculation, but also from a political process; iii) the benefit of setting water abstraction charges at basin or catchment level, in conjunction with robust, action-oriented river basin management plans; iv) how abstraction charges operate, not in isolation, but *in combination with* water entitlements; and v) the use of proxies, which are as effective and more practical than detailed calculations.

Water abstraction charges are commonly managed at sub-national levels. Among other examples, water charges are administered at regional level in Belgium, provincial level in Canada, and at state level (Land) in Germany.

Central or federal governments can provide guidance on incentives. In Canada, the Canadian Council for Ministers of the Environment (CCME) released a guidance document for water valuation in 2010 (available on line). The document concludes that a successful application of water valuation should account for two further aspects, in addition to practical issues: best practice in economic analysis, and stakeholder engagement. In addition, in June 2015, CCME identified a set of options to guide a jurisdiction if it chooses to develop a water pricing framework for industrial or agricultural uses (available on line).

The level of the water charge is usually differentiated by water source (groundwater or surface water) and by the type of users (residential, industry, agriculture). Specific tax rates are sometimes applied to special zones (e.g. Water Apportionment Areas in France) or specific aquifers/rivers (e.g. aquifer-specific abstraction charges are applied in the Flanders region in Belgium and in Estonia). Higher charges are often imposed on groundwater than on surface water (one exception is the Czech Republic, see below). They primarily target households and industries, while the agricultural sector commonly benefits from lower rates or from exemptions.

The objective of the charge is not always stated, in particular regarding its expected achievement in environmental terms. One exception is the legal text introducing the water abstraction charges in Baden-Württemberg (one of the German states). The text makes clear that the charge has three main functions: raising awareness through application of the precautionary principle, incentivising users to save water, and re-balancing competitiveness (self-extraction versus supply through networks).

Charges are volumetric in most cases, with the user paying a unitary rate per cubic metre abstracted (e.g. France). Other structures include fixed charges per hectare for non-metered agricultural abstraction or a price per megawatt-hour for energy production. In Belgium (Flanders region), the level of the tax varies with the quantity of water abstracted. For groundwater abstraction, water abstracted is sometimes paid through increasing block tariff (IBT): the volumetric price is higher for larger volumes. In contrast, the volumetric price sometimes decreases for larger volumes in the case of surface water, the so-called decreasing block tariff (DBT). One issue with decreasing or increasing block tariffs is that the market signal sent to consumers on the value of the resource might become misleading.

Some countries (such as the Netherlands and Denmark) apply a water consumption tax, which is not directly based on abstraction. In Denmark, the tax is equivalent to EUR 823 per 1 000 m^3. In Japan, water pricing arrangements basically do not reflect water scarcity. Only legal and authoritative instrument such as use restriction is implemented during episodes of scarcity to co-ordinate water use among stakeholders (OECD, 2015b). However, abstraction charges vary according to each prefecture and its ordinance, and some prefectures (Kanagawa-ken, Okayama-ken, Tottori-ken, Akita-ken, Kumamoto-ken, Yamaguchi-ken, Shimane-ken, Toyama-ken, among 47 prefectures) have collected local level "Water Conservation Tax" (Lee, 2009).[10] This tax is set to cover environmental costs (mainly cost generated from groundwater abstraction) and pollution charges.

A low natural level of water supply causes water stress in Israel, an absolute water scarce country according to the Falkenmark indicator. This stress has been mitigated by active policies like the recycling of water (75% of wastewater being treated and used for agriculture, industry, etc., replacing the use of drinking water) and the investments in water infrastructure like desalination plants (expected to provide one third of total water demand in Israel by 2020). Water abstraction is regulated by the volume that can be abstracted and by whom, and the basis for the charge is the volume abstracted. The principle for the water abstraction charges is cost recovery, i.e. tariffs should reflect the total supply cost, including capital expenses, energy expenses, and expenses related to operation and maintenance. The method to include a component that reflects scarcity in the tariffs is to add the expenses related to the purchasing of desalinated water on top of the other expenses. Water desalination is the most expensive way of providing water in Israel and therefore reflects the marginal cost of producing water which is the argument why it is included to reflect scarcity.

The Czech Republic and the Netherlands are two examples that can illustrate the importance of a well-designed tax to set the right incentives. In the Czech Republic higher charges are imposed on surface water: almost two to three times higher than for groundwater abstraction. This led to a significant increase in water abstraction from aquifers as a substitute for surface water (Ministry of Agriculture of the Czech Republic, 2015; Reynaud, 2015). Hence, the level and the basis for the tax will influence abstractors' behaviour.

The Netherlands implemented a national groundwater abstraction tax in 1995 that was then repealed in 2012 due to competitiveness concerns in the context of the economic crisis and pressures from the water industry.[11] The tax applied to the abstraction of groundwater by water utilities (water companies then passed the tax on to their customers' bills) and other entities (industry; agriculture). Some exemptions applied, such as sprinkling and irrigating land if less than 40 000 m^3 per year were extracted, and pumps with a capacity lower than 10 m^3 per hour, implying that in practice agriculture was more or less completely exempted. Only groundwater abstraction was charged since the goal of this policy was to increase the price of groundwater so that surface water would be used instead. The tax was administered by the Ministry of Finance and the Central Environmental Tax Unit.

The exemption system had unintended (and negative) consequences: the pumping capacity exemption created an incentive for farmers to use several small pumps to avoid paying the tax, which was said to result in an overexploitation of groundwater. The system also raised complaints from water-intensive industries since the groundwater tax was different for industries supplied by water companies and for those with self-extraction, hence raising competitiveness issues. Since 2012 when the central government's environmental tax on groundwater abstractions was abolished,[12] only the provincial groundwater tax remains in place, serving as a financial source of regional groundwater management (EUR 15.2 million in 2012) (OECD, 2014b). However, the amount of water available for consumptive use is limited according to a statutory instrument called the Peil belsuit and the country invokes a system of priority uses during prolonged drought.[13]

In Portugal, a water resources tax was set by a law approved in 2008. It applies to all private water users abstracting water. The tax implements the idea that private users of water resources should compensate the cost generated to the community, or restore the benefits to the community (User Pays and Polluter Pays principles). The tax aims at collecting funds for water management. It also aims at guiding private users towards a more efficient use of water and to allocate water where it creates most value. The water resource tax sets off i) the advantage resulting from the private use of public water; ii) the environmental costs related to the activities that affect water resources; and iii) the administrative costs of water management. The components of the tax reflect the varied shortage of water resources in different parts of the territory (Souza d'Alte, 2010). At the beginning of the implementation process, water utilities, hydropower generators, paper industries and irrigators emerged as the largest contributors.

The private sector provides another illustration of how a water abstraction charge can be used to drive decisions about location of activities and investments. Some corporations, in particular in the agro-food industry, invest in water management, to minimise their production cost now and in the future, to engage in sustainable manufacturing, and to enhance their licence to operate (by minimising tensions with other water users where they operate). Nestlé introduced an *internal shadow price of water* in all countries where it operates and where water is not charged. The absence of a charge should not mean that water has no value, and Nestlé endeavours to calculate and reflect that value: it sets shadow prices for internal decisions (e.g. investment decision on the location of a new factory, or on the introduction of water saving technologies), based on water availability in the region, using a very gross proxy. In agricultural areas where water is abundant, a price of USD 1/m^3 was used; in water scarce areas, the price was USD 5/m^3. That shadow price directly affected business decisions, of locating activities and production sites, as it helped make the economic cases for investment in water efficiency. An important message is that the basis of the calculation does not need to be sophisticated: a rough proxy can be used, which sends the right signal to decision makers.

Table 3.5. **Water abstraction charges in selected countries**

	Bases for charge			Level of the tax		
Country	User	Groundwater (GW) or Surface Water (SW)	Other	m³/ha/mWh	(EUR/1 000 m³)	Authority in charge/ Levied by
Australia (Australia Capital Territory)	yes, urban water supply versus others	Not differentiated		m³	Urban water supply: EUR 360 (2013) Others: EUR 175 (2013)	ACT Government
Belgium (Flanders)	yes	GW	Varies by aquifer	m³ + IBT	500 – 30 000 m³: EUR 50 above 30 000 m³: EUR 62	Regional authority
	yes	SW		m³ + DBT	<1 Mm³: EUR 63	Regional authority
Belgium (Wallonia)	yes	GW only		m³ + IBT	3 000-20 000 m³: EUR 25 20 000 – 100 000 m³: EUR 50 > 100 000 m³: EUR 74	Regional authority
Czech Republic	yes	SW	Varies by river basin	m³	country average: EUR 155 (2014)	River boards
	yes	GW		m³	< 6 000 m³ per year: exempted country average: EUR 35 (2013)	
Estonia	Yes (but few categories)	GW, SW and mineral water	Varies by aquifer	m³	GW: 60 – EUR 160 SW: 2 – EUR 38	
France	yes	Differentiated between GW and SW	Water Apportionment Area (ZRE)	m³ if metered; proxied or fixed fee if not metered	Drinking water supply: EUR 33-51 (2009) Irrigation: EUR 2-16 (2009)[1] Hydropower: EUR 0.17-0.64 (2009)	Water Agencies (hydrographic basin level)
Germany (Baden-Württemberg)	no	GW			Public water supply, heat production and others (incl. irrigation): EUR 51	State government
	yes	SW			Public water supply: EUR 51 Heat production and others (not incl. irrigation): EUR 10	State government
Hungary	yes	Differentiated according to the type of water (e.g. surface water, thermal water)		m³ or self-estimation for irrigation	EUR 0-100	Regional Water Directorate
Poland	yes	GW		m³	EUR 15-25/1 000 m³ (2011)	
	yes	SW		m³	EUR 9-13/1 000 m³ (2011)	
Slovenia	yes, large number of categories			m³ except for energy production and hydropower (mWh)	Drinking water supply: EUR 55 Irrigation of agricultural land: EUR 1 Cooling: EUR 4 Hydropower plants above 10 MW: EUR 1 500 Hydropower plants below 10 MW: EUR 190	

Note: 1. Tax applied to non-gravitational irrigation.

Source: OECD database on economic instruments (no date provided).

3. Options for reform

The previous sections have argued that Korea would benefit from a reform of the three economic instruments under MoLIT's authority to manage water efficiency. This section explores how they could be reformed and puts forth a number of actions that can facilitate reform. The emphasis is on the reform of existing instruments with the view to fix some of their limitations and to exploit their full potential. Possible measures relate to both the revenue raising capacity and the efficiency of expenditures financed by these instruments.

The section also proposes that an abstraction charge be considered in the longer term. Should that option be considered, the abstraction charge should operate in combination with other instruments, in particular a zoning of water scarce regions and water entitlements (see Chapter 5). The charge would be more appropriate if it considers conditions for relative water scarcity or how water supply meets demand. Cognisant that this suggestion is quite distant from prevailing practice in Korea, This report suggests some measures to manage transition toward the implementation of an abstraction charge, including the consideration of a shadow opportunity cost of using water. A shadow opportunity cost would help to i) simulate the potential impact of an abstraction charge on decisions by cities and corporates, ii) stimulate investment in water saving technologies and practices, and iii) drive decisions about supply augmentation, infrastructure design and financing.

Figure 3.5 suggests a tentative sequence of action to develop economic instruments that contribute to water use efficiency in Korea.

Figure 3.5. **Strengthening EPIs for water use efficiency in Korea**

A staged approach

Upgrade the river water use fee	• Apply the fee to the volume of entitlement • Mandate municipalities to report on revenues • Mandate municipalities to develop expenditure programmes financed by revenues • Assess expenditure programmes, with a view to enhance their contribution to water security
Upgrade dam water tariff and multiregional water tariff	• Set acceptable levels of security for water supply • Mandate K-water to develop expenditure programmes at basin or catchment level to achieve the set level of security • Mandate the economic regulator to review expenditure programmes • Endorse the financial strategy (combination of budget transfers, revenues from tariffs and other sources of finance
Consider abstraction charges as a module in existing instruments	• Define conditions where an abstraction charge would apply (zones of structural water deficit, definition of a drought episode) • Set proxies for opportunity costs and environmental costs of using water • Set shadow abstraction charges that reflect local conditions • Mandate water users and K-water to use the shadow abstraction charge when applying for a water licence or considering supply augmentation measures
Take accompanying measures	• Set a water monitoring system at regional level • Develop a regional monetary water accounting system • Develop an operational methodology to compute opportunity costs and environmental costs • Monitor the social consequences and design targeted social measures • Design interregional budgetary transfers to level out benefits from past public support to water supply augmentation

3.1. Strengthen economic policy instruments

Korean authorities increasingly realise that supply augmentation can face difficulties and are looking for alternatives. It is noteworthy that previous attempts to promote water use efficiency, including the 2000 Comprehensive Plan for Water Savings and the 2007 Comprehensive Plan on National Water Demand Management, have resulted in significant water savings, e.g. 930 million tonnes of tap water saved by 2012 for the 2000 Plan and a target of 1 021 million tonnes of water saved by the end of 2016 for the 2007 Plan. However, these achievements essentially relied on technical fixes (expansion of water services, replacement of ageing infrastructure), which are costly responses to current and future challenges. The question remains whether these savings are sufficient and could have been achieved at a lesser cost through a combination of policy instruments, including economic ones. The development of the Smart Water Management in Korea (see Chapter 4) confirms that Korean authorities are looking for alternative options to address water scarcity and flood risks and make the best use of available water and assets, while minimising or postponing investment in additional infrastructure to augment supply.

Even if water access is not currently an issue in Korea under normal conditions, it will become a pressing policy issue in the future due to, *inter alia*, continued economic growth and urbanisation, and climate change. Economic instruments can play an important role and complement the technical fixes used so far in Korean water management. When appropriately set, water tariffs and abstraction charges can inform abstractors about the value of water and provide incentives to use water efficiently, including in episodes of droughts. They can also provide incentives to direct economic and urban developments in regions where water is more abundant, avoiding creating additional liabilities as regards water security (See Box 3.5). Well-designed water tariffs and abstraction charges can also stimulate innovation and private sector investments (in water-saving technologies and practices or in water storage) and allocate water where it creates more (social, environmental and economic) value.

Box 3.5. **The tale of a Korean brewer**

A brewer – a water intensive industry – considers setting up a new brewery in Korea. All other things being equal, it makes economic and environmental sense to locate the new facility in an area where water is abundant and competition to access it is limited. Under these circumstances, the brewery could operate without affecting other water users and new capacities to augment water supply would not be needed. If the brewer decides to locate the brewery in a region where water is scarce and competition to access it is high, two options will have to be considered: either reduce the volume of water available for other users, potentially affecting the revenue they generate or their ability to operate, or augment supply by building a new storage capacity. The first comes at a cost for the economy (reduced output from other industries), the environment (if environmental flows are affected) or the community (if households access to water is hindered). The second also comes at a cost: the cost of building additional infrastructure to augment supply; that cost is potentially born by the community, when the infrastructure is financed through public budgets. The costs incurred by the community result from the brewer's individual decision; therefore, They should (at least partially) be covered by the brewer. Such is the rationale of setting up an abstraction charge that reflects the opportunity cost of using water for that brewery.

Box 3.5. **The tale of a Korean brewer** *(continued)*

The same example suggests that abstraction charges should reflect local conditions. Should the abstraction charge be set nationally, the brewer would have no incentive to locate in a particular region. It would merely decide whether or not to operate a new brewery. If the abstraction charge reflects the opportunity cost of water in regions where water is abundant (i.e. the opportunity cost will be low), the investment is likely to materialise and could be located anywhere, including in water scarce regions, generating a cost that is not covered by the charge. If the abstraction charge reflects the opportunity cost of water-scarce regions, water availability will be protected but water abundant regions will lose the opportunity of a new activity.

3.2. Strengthen existing instruments

The analyses in this chapter have established that existing tariffs or charges to manage water quantity could better contribute to water policy objectives.

Revenues collected through the *river water use fee* should apply to entitlements and be monitored; their contribution to water security in the basin (essentially protection against risks of floods or scarcity) should be assessed. This measure would be an incentive for local authorities to disclose information on water use and water expenditures. It would also be an incentive to revise water entitlements so that over time they reflect actual use, thereby generating more flexibility to reallocate available water, possibly to new comers, in basins where water was over-allocated but not over-used. Expenditure programmes financed by these revenues should be assessed by MoLIT. Once the efficiency of this instrument is enhanced, rates can be adjusted to match financing needs.

Dam water tariffs and *multi-regional water tariffs* should be defined so as to cover the cost of service provision, and ensure maximum value of expenditures financed from these revenues. A price ceiling could be considered similar to the economic regulation of water supply and sanitation services in England and Wales. MoLIT could set objectives as regards infrastructure development and service provision and define an acceptable level of water risk as regards water supply and flood control (see OECD 2013, on the concept of acceptable level of water-related risks). K-water could develop investment programmes to achieve these objectives and propose a financing strategy to cover the costs of such programmes, which could include a revision of water tariffs. Under this scenario, an economic regulator would review the investment programmes, their opportunity and their efficiency to validate the financing strategy and the proposed revision of water tariffs. The economic regulator and K-water would negotiate an agreement on the investment programmes and the appropriate financial strategy, in particular the combination of budget support (if any) and revenues from Dam water tariffs and multi-regional water tariffs.

From an efficiency perspective, it would be preferable if the costs of building and operating a particular infrastructure (dam of multi-regional water supply system) could be known and singled out. Even if tariffs are set nationally, knowledge of capital and operating costs per infrastructure can inform decisions about the opportunity of building new infrastructures; it can document the hidden budgetary transfers from basins or regions where supplying water comes cheap to others where the cost of supplying water is high. MoLIT and K-water would benefit from a review of international good practice for sharing disaggregated information about the cost of supplying dam or bulk water to different parts of the country.

In addition, the relative costs of securing access to water should feature on the water bill. This could be done by singling out river water use fee, dam water charges or multi-regional water charges, or an abstraction charge in the water bill, for industry and domestic users. This would also be an incentive for local authorities to reflect any increase in these economic instruments in the water bill, so that tariff policy eventually affects users' behaviour.

3.3. Consider a water abstraction charge

A proper water abstraction charge can contribute to several policy objectives in Korea. It can contribute to allocating water where it creates most value and discourage inefficient water use. It can also stimulate investment in supply augmentation, when appropriate.

Should Korean authorities consider setting up a water abstraction charge, the objective of the charge should be clearly stated; its design should be transparent, as well as the use of the revenues collected. Authorities in charge of the management of the water abstraction charge would make a clear statement of what this charge is meant to achieve: protecting the water resources by providing the right incentives to all water users. In order for the charge to promote efficiency, it should reflect the true full social cost, including environmental externalities and the opportunity cost of using water (see Figure 3.3), using some of the methods suggested in the chapter. Currently, as described earlier in this chapter, dam and multi-regional water tariffs in Korea are set to only cover O&M and partial capital charges; river water use fee essentially covers administrative costs. None of them operate as an abstraction charge.

In principle, an abstraction charge should reflect water availability and usage, scarcity risks, and seasonal variability. These features differ across Korea's six hydrographic zones; therefore, water users should face signals reflecting these different conditions. An abstraction charge that is adjusted in the six hydrographical zones would contribute to this objective.

Box 3.6. **The rationale for water abstraction charges that reflect local conditions**

Abstraction charges are best differentiated by hydrographic zones in order for each category of users to receive an appropriate signal on the value of the resource (economic efficiency objective), and for equity reasons.

Abstraction charges that are the same across the entire country are unfair. For example, take a country divided into two regions, one vulnerable region in which the opportunity cost of using water is estimated at KRW 2000 per cubic metre abstracted, and one less vulnerable region where the opportunity cost is estimated at KRW 500 per cubic metre abstracted. Economic efficiency requires that the abstraction charge be set at KRW 2000 per cubic metre in the vulnerable region and at KRW 500 per cubic metre in the less vulnerable region. With such a differentiated abstraction charge, users in each region will face the right price signal and adjust their water use optimally so that the overall welfare of the two regions reaches its maximum and the water resources are not overexploited. If policy makers decide to set an abstraction charge that is the same in the two regions, then the charge has to be set at KRW 2000 per cubic metre in order to preserve the water resource. Indeed, if the abstraction charge is set below KRW 2000 per cubic metre, then users in the vulnerable region will abstract too much water (because the price is too low for them to get the right signal) and water resources will be overexploited. But if the abstraction charge is set at KRW 2000 per cubic metre for everyone, users in the less vulnerable region will reduce their consumption below the optimal level (since they will pay a too high price for water). This could result in a loss in welfare for households and a loss in value added for industrial and agricultural users.

Box 3.6. **The rationale for water abstraction charges that reflect local conditions** *(continued)*

Moreover, charges may also have to be differentiated by areas to adequately address environmental externalities. Here a differentiation based on hydrographic zones may not be sufficient since there may be local (within hydrographic zones) variation in the environmental effects from water abstraction. For instance, the need for instream protection, as well as other environmental externalities, is likely to vary within each hydrographic zone. Further abstraction in one area may have minor or no effects on aquatic life or recreational activities, whereas the effects may be substantial in other areas. Hence, a differentiation at zone level will not adequately internalise the environmental impact, and therefore charges should be differentiated to reflect this variation.

In addition, because of precipitation patterns in Korea, water abstraction charges could vary according to seasons: water abstracted in the summer does not have the same value as the same volume of water abstracted in the winter. Reflecting such changes in the value of water (environmental or opportunity costs of using water) can drive private investment in water storage. An example from the UK (see Annex 3.A2) provides some guidance on how to incorporate such variations in the value of water. In practice, most OECD countries consider that seasonal variations in abstraction charges add a lot of complexities and seldom use them. For the same reason, it makes sense to only consider seasonal variation at a later stage in Korea.

Setting abstraction charges at local (basin) level, while common practice in OECD countries, raises special concerns in Korea given the historic rationale behind investment in water infrastructure, which was meant to even out conditions of access to water across the country (see Box 3.7). Section 3.4 discusses how these concerns can be addressed and provides recommendations on how to transition from the existing instruments managed by MoLIT to abstraction charges that contribute to water policy objectives in Korea.

Box 3.7. **Abstraction charges and equity issues in Korea**

Economic theory and good international practice recommend that abstraction charges reflect the opportunity costs of using water. This is an issue in Korea, where the development of water infrastructures (dams and canals) was essentially meant to drive economic development and to provide equal opportunities to all parts of the national territory to develop.

The abstraction charge has to be set such that each user gets the right signal on the cost it imposes on other users and on the environment; it is meant to guide future behaviour and investment. Considerations that some hydrographic regions have benefitted from investments made in the past to secure water supply only blur the message.

Investments in water supply in Korea should be evaluated using benefit-cost analysis (BCA) to ensure that new investments (in new or old infrastructure) improve welfare. If financed by the national government, the BCA should take a national perspective. This may lead to investments not being evenly distributed to all regions. To address this uneven distribution of resources from the national government, interregional budgetary transfers – not abstraction charges – are the appropriate tool to acknowledge that some regions benefited more than others from investments made in the past to secure water resources.

Box 3.7. **Abstraction charges and equity issues in Korea** *(continued)*

The abstraction charge should be implemented to price the water according to the cost of providing water (user costs), the opportunity cost of water (scarcity) and environmental externalities only. Fairness or equity issues have to be addressed using an instrument that is disconnected from the water charge in order not to distort the signal sent to water users.

3.4. Manage the transition towards abstraction charges

Should abstraction charges be considered in Korea, the following steps could be considered to ensure a smooth transition:

1. Amend the three instruments under review along the lines suggested in the sections above.
2. Supplement each of the three instruments with a module that reflects the opportunity costs of using water. That module would only operate under special conditions characterised as drought. These conditions could be defined at national level, for instance by the Water Management Committee. These conditions could be permanent (equivalent to the French *Zone de Répartition des Eaux*), or time-bound (in the case of a drought episode affecting a region which does not usually face scarcity).
3. Gradually differentiate opportunity costs by region to reap the full benefit of the instrument. In parallel, the definition of special conditions under which the opportunity cost module would come into play could be devolved to river basin organisations, for instance the Flood Control Offices.

Properly designed and managed abstraction charges are quite distant from prevailing uses of economic instruments to manage water quantity in Korea; therefore, the opportunity of introducing them should be further analysed. A range of government agencies (Ministry of Environment, MAFRA, Ministry of the Interior, Ministry of Strategy and Finance…) would need to engage in the conversation, to ensure policy coherence and to provide technical support (in assessing environmental and opportunity costs, for instance). Engaging stakeholders at national and local levels is essential, while providing them with robust economic analysis of the (implicit) budgetary, economic and social transfers associated with the three instruments under review.

Any attempt to introduce abstraction charges should benefit from regular reviews and assessments. Equity issues deserve particular attention, to ensure that no particular basin, industry or water user can abuse from a privileged condition.

A shadow abstraction charge could be considered to raise awareness and promote understanding of abstraction charges, to simulate the potential consequences of such an instrument, and to gain experience with using it. It would mimic a genuine abstraction charge, but would not be collected in practice. The shadow abstraction charge would be based on proxies and would reflect local conditions at basin or catchment level. It would be used by authorities and water users to guide decisions regarding water licensing, investment in water use efficiency, and the design and financing of infrastructures to augment supply. After a transition period, the shadow abstraction charge could be revised and fine-tuned to reflect lessons from experience, and then turned into a real abstraction charge that would apply to water users.

In parallel, the following actions are required to deliver the information needed to support and upgrade existing economic instruments. They would benefit from inter-governmental co-operation:

1. ***Set a water monitoring system (water accounting system) at regional level.*** This water monitoring system (or water accounting system) should at least measure for each region: water withdrawals, water consumption and water discharge by type of user (agriculture, industry, household, and environment) and type of water resource (surface water, groundwater, and dam water). This water monitoring system could be decentralised to each region provided the process is carried out across regions using the same method, i.e. the same technical specifications.
2. ***Develop a regional monetary water accounting system.*** Set an accounting system that measures the cost of water supply (O&M cost, capital cost) for each region: cost of water supply per type of water resource (surface water, groundwater, and dam water) and per user type (agriculture, industry, household). Define clear rules for amortisation of capital. Note that such an accounting system is also required to enhance the capacity of river water use fee, dam and multi-regional water tariffs to perform as financing instruments, and to ensure that those who generate costs for the community cover these costs (Beneficiary or User Pays principle).
3. ***Develop an operational methodology for imputing external costs of using water to water users.*** Ideally, this step consists of measuring and informing water users about the opportunity cost of using water and the monetary value of externalities. This requires very substantial work in terms of data collection and data processing. As a practical alternative, a "zoning approach" could be used and would consist in: i) identifying within each river basin zone of specific interest from a water management point of view (highly polluted areas, areas with water scarcity, areas with particular environmental characteristics), and ii) specifying water resources across zones.

A water abstraction charge could be introduced at national level when the comprehensive set of information is collected and analysed.[14] The abstraction charge at national level should reflect the highest opportunity cost in the country; otherwise, it will not be able to address issues of scarcity where they are most severe (see Box 3.6 above). In addition, disaggregated information could be used to inform policy and investment decisions related to land use and infrastructure development.

The main characteristics of the proposed abstraction charges are:

- The abstraction charges would be based on the volume abstracted over a year. If the instrument is used to provide incentives to water users, then the charge for abstracting water has to be linked to quantities abstracted (or used). Only in cases where metering is not available (e.g. because too costly) should other bases (e.g. production output), or other pricing systems (fixed fee for example) be chosen.
- Abstraction charges should be universal and paid by all sectors. Water charges may vary across user type (household, industry, agriculture) to reflect the variation in the consumptive use portion of the abstracted water between type of uses (cf. the case of England and Wales). Return flows need to be specified by sector. This can only be done in consultation with relevant stakeholders, at basin or catchment level, to reflect local practices.
- Implementation of water charges could be incremental, including a transition period, but a clear target must be set from the beginning.[15] Possibilities to revise the level of the charge should be included from the beginning.

- The social impacts of abstraction charges should be monitored. This requires an information system to monitor the impacts of abstraction charges to allow Korean authorities to assess the impacts of the water charges on different users (e.g. households and industries), and to evaluate their performance to provide incentives and to allocate water resources efficiently. An example of such an information system can be found in Australia. The Australia water accounts include a financial section, which presents monetary data for supply and use of water within the Australian economy (see Box 3.8). Again, the three instruments reviewed in this chapter, represent a minimal share of the water bill covered by final users. Therefore, it is very unlikely that any revision of these instruments or the introduction of an abstraction charge will trigger severe affordability issues. Should affordability issues emerge from the monitoring system, mechanisms could be implemented, e.g. a social fund dedicated to helping the poor to pay their water bills.

Box 3.8. **Examples of national and regional water accounts**

Water accounts provide a comprehensive picture of water resources management and water use. They are usually produced at national level, but also in some cases at regional levels. Water accounts disclose information about water stocks and flows, water abstraction and uses for economic, social, cultural and environmental benefit.

A good example of water accounts is Australia (www.abs.gov.au/ausstats/abs@.nsf/mf/4610.0). Since 2010, Australia has released annual national water accounts and regional water accounts for ten nationally significant water use regions (e.g. Adelaide, Burdekin, Canberra, Daly, Melbourne, Murray-Darling Basin, Ord, Perth, South-East Queensland and Sydney). Accounts are under the responsibility of the Australian Bureau of Meteorology.

Another example of a country that has developed a good water account system (national level only) is Denmark (www.dst.dk/en/Statistik/emner/groent-nationalregnskab/vand-og-spildevand#). Water use information (by type of water user and by source of water) has been provided on an annual basis since 2010.

Initiatives to harmonise water accounts have been conducted at an international level; see the Water Assets Accounts initiative in Europe.

Sources: Water Account Australia (2016), *Water Account, Australia, 2014-15*; European Environment Agency (2013), *Results and lessons from implementing the Water Assets Accounts in the EEA area*.

The authority in charge of defining a water abstraction charge should be independent and rules to define a water abstraction charge should be transparent. The agency should rely on expertise and advice involving scientists and stakeholders (farmers, municipality, industrial and energy producers, environmental defence and consumer protection NGOs, etc.). It could be organised through water committees such as the *Comités de bassin* in France.

Different water users should be treated equally to provide a sense of fairness. Differences in water charges should be based only on water-use characteristics (volume, return flow, location) and the likely impact of water abstraction on the environmental and opportunity costs. No exemptions should be made, including reduced charges for essential uses such as drinking water.

The development of an abstraction charge should be accompanied by an outreach campaign that explains to water users i) the rationale for the water abstraction charge, ii) what the revenues will be funding, and iii) the resulting improvements in water security.

A reform of the three instruments reviewed in this chapter should be combined with a comprehensive review of economic instruments for water management in Korea (including the ones under the remit of other ministries) and a reform of water allocation regimes. Pricing instruments contribute to allocating water where it creates most value. Sensible water prices are particularly valuable, when episodes of scarcity are increasingly frequent. They may not be sufficient to address severe droughts. Allocation regimes contribute to similar policy objectives in normal times and in times of scarcity. A smart combination of both sets of instruments can contribute to water security at least cost for society.

3.5. Adjust the institutional framework

The design and implementation of economic instruments that contribute to water use efficiency in Korea require institutional arrangements that reflect local conditions (at least at basin level) and engage with relevant stakeholders at the right scale (national, basin, or else). The sections above signalled the need to produce and communicate new information on water availability and use, revenues generated and expenditure programmes, and improvements in water security. They alluded to the need to set acceptable levels of security for water supply, an inherently political process.

In addition, there is some value in empowering river basin organisations. One option might be to include basin organisations or authorities that represent the basin in the constituency of the Water Tariff Committee. This could be considered in the broader context of valuing river basin management in Korea, an issue which reaches beyond the scope of this policy dialogue. Should that option be considered, it may be appropriate to assess capacity needs and allocate human and technical resources to basin organisations, so that they can deliver on their new tasks and responsibilities. The point is not so much about devolving responsibilities about setting abstraction charges to river basin organisations, but rather about building partnerships between different levels of water governance.

Moreover, to better serve the purpose of the Water Tariff Committee, rules to appoint members could be improved with a view to facilitate access of distinct perspectives (including regional ones). More could be done to enhance information sharing and provide feedback on the content of the reviews. MoLIT has achieved progress in stakeholder engagement when it concerns its approach to infrastructure projects, and K-water is raising the profile of stakeholders in its future strategy. There are benefits to extending the same attention and efforts to economic instruments for water quantity management. Indeed, encouraging user engagement is a critical step toward building social acceptance and secure buy-in for policy choices. It also helps to raise awareness about current and future risks of too much or too little water, and to justify the need to transition to more sustainable practices and regimes. It can finally strengthen the willingness of users to pay for water.

To strengthen stakeholder engagement in tariff-setting processes, three areas of action, which could be implemented in sequence, can be considered:

- Central and local authorities and public institutions could organise partnerships and invest in educational programmes and information campaigns on the real cost of water. The campaigns would explain how water charges are calculated and revenues are spent. This would help to build a broad public understanding of water charges and tariffs for water supply and sanitation services, water-related risks and the costs of risk mitigation policies. It would raise appetite for consumers to get involved.

- **New opportunities for stakeholder engagement could be envisaged at basin level.** River basin organisations under MoLIT's responsibility could be tasked to set up platforms for water users to share their concerns and influence policy choices. By participating in more regular and substantial interaction and co-operation with public authorities, citizens and civil society organisations would contribute to enhanced transparency, responsiveness, and accountability of water policies, thus improving the quality of decision-making processes on economic instruments.
- A strategic management approach could be adopted, including a clear vision and goals of the programme and a result-oriented performance management system of the programme. For example, values and goals of such an engagement programme could be increasing social learning, citizen empowerment for decision-making, transparency or trust in government. The programme could also utilise information technologies and emerging social media tools to activate greater involvement in water pricing policy formulation and to create greater information exchange between citizens and the government. MoLIT, K-water and basin organisations, together with NGOs and consumer associations, have a critical role to play in setting the enabling environment for effective and outcome-oriented engagement, while maintaining a balanced combination between government support and bottom-up efforts.

Annex 3.A1

Water abstraction charges in practice – France

Under normal conditions, France has enough water resources to satisfy the demand of all users and to maintain the environment in good ecological condition. However, because some areas may occasionally suffer from scarcity issues and conflicts of use, a charge on water abstraction has been implemented in order to make users internalise opportunity costs of using water. Opportunity costs, as well as pollution costs, have to be covered by water users through the implementation of an incentive-based charge in accordance with the European Water Framework Directive 2000/60/EC which was transposed into the French law in 2004.

France (mainland) is made of seven hydrographic basins (Adour-Garonne, Artois-Picardie, Corse, Loire-Bretagne, Rhin-Meuse, Rhône-Mediterranée and Seine-Normandie) managed by six water agencies. The charge on water abstraction is the main tool used for water quantity management in France. In periods of acute scarcity, it can be complemented by regulatory instruments such as restrictions that apply to specific water uses.

General principles

The rate of the water abstraction charge is defined at the hydrographic basin level but cannot be set above the national ceiling defined by French national law for major uses (see below for greater details). The law also sets principles regarding the basis for charging as well as exemptions.

The charge is a component of the price of water that is paid by all users: households, industries, and farmers. It is collected by water agencies within each hydrographic basin. The charge builds on a characterisation of water resources as regards availability: Category 1 and Category 2 resources are those situated respectively inside and outside Water Apportionment Areas (ZREs in French, for *Zones de Répartition des Eaux*), which are zones characterised by a chronic water deficit. The spatial distinction between Category 1 and Category 2 resources reflects the higher opportunity cost in ZREs where there is a serious imbalance between water demand and water availability. There is no adjustment of the charge across seasons, even if the pressure on the resource is usually higher during the summer.

Principles are set by the national law:

- Basis: The charge is based on the volume abstracted over a year. Anyone using a well or borehole is required to install a water meter. If abstracted water is not measured, a fee applies. The latter is based on estimates of volumes abstracted or measures from representative samples. For example, water used for gravitational irrigation is taxed on the basis of 10 000 m^3 per irrigated hectare.

- The water agency in each hydrographic basin sets abstracted volumes below which users are exempted from the charge. These volumes cannot be larger than 10 000 m^3 per year for abstraction from Category 1 resources and 7 000 m^3 for abstraction from Category 2 resources.
- Exemptions: Water abstracted from the sea; water used for mines which are no longer active; for aquaculture; for geothermic activities; for restoring the environment (if those volumes are abstracted outside the period of low water flows); water abstracted to protect perennial crops against frost.
- The charge is set by each water agency (in EUR per 1 000 m^3) within the limit of the following ceilings.
- French legal documents do not provide justification for the level of the tax limit or for the factor 2-difference between Category 1 and Category 2 resources. As far as we know, these limits were not based on any scientific evidence assessing the costs induced by water abstraction. They are proxies that reflect a consensus among stakeholders at a moment in time.
- Regarding abstracted volumes for irrigation purposes: when abstraction is made by a water users association (*Organismes Uniques de Gestion Collective* or OUGC in French, which gathers several individual irrigators) within a Category 2 resource (ZRE), then the tax rate of Category 1 resources applies. This is an incentive for individual farmers to join a water user association and jointly manage a collective entitlement, which is considered a more effective way to allocate water when it is scarce.

Table 3.A1.1. **Water abstraction charge limits in France**

EUR/1 000 m^3

Type of use	Category 1 resources Outside ZREs	Category 2 resources Within ZREs
Irrigation (except for gravitational irr.)	36.0	72.0
Gravitational irrigation	5.0	10.0
Drinking water supply	72.0	144.0
Water cooling (if more than 99% of the volume abstracted is returned to the environment)	5.0	10.0
Supplying canals	0.3	0.6
Other economic uses	54.0	108.0

Source: International Office of Water (2009), *Organization of Water Management in France*, Paris.

The level of the tax is then set within each water agency by the River Basin Committee (RBC) (see Table 3.A1.2). RBCs have been created in 1964. They are deliberating bodies that unite all stakeholders from each river basin district. Their composition is as follows: elected government officials (40%), water users (40%), and State representatives (20%). RBCs determine the strategy for the protection of water and aquatic environments for each river basin (the management plans for river basin districts – SDAGE). They vote on the water agency's action programme and the rate of fees, within the limit of the rates set by law.

Table 3.A1.2. **Water abstraction charge rates in France**

2016

	Drinking water supply (EUR/1 000 m³)			Non-gravitational irrigation (EUR/1 000 m³)			
Water agency	groundwater outside ZREs	surface water outside ZREs	within ZREs	groundwater outside ZREs	surface water outside ZREs	groundwater within ZREs	surface within ZREs
Rhin-Meuse	52.0	33.2	144.0	4.7	4.7	72.0	72.0
Loire-Bretagne	34.0	34.0	43.3	12.6	12.6	19.0	19.0
Rhône-Méditerranée-Corse	46.6	30.0	68.3	6.8	6.3	12.9	12.5
Adour-Garonne	42.0	42.0	56.0	8.8	8.8	11.7	11.7
Seine-Normandy[a]	B: 60.0 Q: 67.0	B: 38.0 Q: 46.0	82.0	B: 21.0 Q: 27.0	B: 17.0 Q: 20.0	34.0	34.0

Note: a: In Seine-Normandy, three zones have been defined: basic zone, zone of quantitative pressure, and ZRE. Here, B refers to basic zone while Q indicates zone of quantitative pressure.

Sources: Rhin-Meuse Water Agency (www.eau-rhin-meuse.fr/differentes_redevances); Loire-Bretagne Water Agency (www.eau-loire-bretagne.fr/nos_missions/redevances/tx-ressource-10epgm.pdf); Rhône-Méditerranée-Corse (www.eaurmc.fr/aides-et-redevances/redevances-et-primes/prelevement-deau.html); Adour-Garonne Water Agency (www.eau-adour-garonne.fr/fr/quelle-politique-de-l-eau-en-adour-garonne/un-outil-le-programme-d-intervention-de-l-agence/les-redevances-percues-par-l-agence.html); Seine-Normandy Water Agency (www.eau-seine-normandie.fr/index.php?id=7865); all accessed on 14 September 2016.

Complementary instruments [16]

Water entitlements

Water users are required to hold a water entitlement to abstract water except for small scale domestic water abstractions (under 1 000 m³/year). Entitlements are defined through a process of impact assessment and public inquiry. Abstraction authorisations are granted by local state representatives. According to the abstracted volumes (above or under local thresholds), they can be submitted either to declaration or to authorisation (with a public inquiry). The ZRE (water deficit areas) thresholds are lower than in other areas.

Entitlements should be issued in conformity with the River Basin Management Plans (*Schémas Directeurs d'Aménagement et de Gestion des Eaux*, SDAGE) and the national framework (respect of minimum thresholds for the ecosystem use of water). Water entitlements are unbundled from property titles and are granted for a few years to several decades for permanent use like drinking water abstraction. Six month entitlements can be granted for temporary uses (considered as uses without dramatic environmental impacts).

Prevention and planning instruments

Long-term planning instruments at the hydrographic basin level (SAGE/SDAGE: Water Development and Management Plans) are aimed at preventing crisis situations by planning offer (i.e. volumes that can be abstracted) and use (demand) over the long-term, and defining priorities in terms of water use. Limits on consumptive use through the volumes that can be abstracted are defined within SDAGEs and SAGEs. They are statutory instruments that must be followed. Abstractable volumes have to be redefined after a few years to consider the latest available scientific data (Water Management Plans are re-defined every six years).

The calculation of abstractable volumes takes into account minimum flow requirements: the minimum environmental flow is defined based on a statistical standard as the monthly flow with a probability of exceedance of 1 in 5. Other minimum flow requirements relating to freshwater include “the minimum biological flow” and the “reserved flow”, based on ecological needs.

Non-consumptive uses (e.g. navigation, hydroelectricity, nuclear electricity) are treated in the same management plan as the consumptive uses, in order to find the correct balance. These uses also require a minimum flow.

Drought decrees (crisis management)

In case of shortage, an “exceptional circumstances status” can be declared by a crisis unit, called the “Drought Unit”. This unit is convened by the prefect (the local state representative) and involves stakeholders. It is mandated to suspend prevailing entitlements and to allocate water according to a set list of priority uses, as defined in the Decrees. The definition of “exceptional circumstances” is based on local indicators (low flow in rivers and the piezometric water level for aquifers). Pre-defined priority classes are as follows: i) Domestic + National Security (drinking water; health-related issues; civil safety, including cooling of nuclear power plants); ii) Environment (a balance between ecosystems and economic uses); and iii) Agriculture, Industrial, Energy Production and Transfer to the Sea or Another System.

Monitoring and enforcement

The Water Police (local representatives of central government administration) is responsible for monitoring and enforcement. For those not complying with the rules, there are two kinds of sanctions: administrative sanctions (from the formal notice to the suspension of the authorisation) and penalties, with fines (the fine can amount to EUR 1 500).

Limitations of the system [17]

The incentive power of the abstraction charge remains limited (for households in particular) since it represents a small share of the average price of water (around 1.3%). In 2013, the average price for water and wastewater sanitation services was EUR 3.85/m^3 while the abstraction tax was around EUR 0.5/m^3 (Observatoire des services publics d’eau et d’assainissement, 2012, Onema, 2015).

There is currently no reference value for the opportunity costs on which tax rates on abstraction could be based. It is therefore difficult to judge whether the level of the charge is appropriate. However, opportunity costs should be related to the pressure exerted on the resource by the various users and higher rates should be imposed on uses that exert greater pressure. This principle does not seem to apply in general. For example, abstraction charge rates are higher for domestic users while volumes abstracted for drinking water supply are much lower than volumes abstracted for irrigation purposes. The pressure exerted on the resource can also be assessed through the difference between the volumes of water that are abstracted and the volumes that are returned to the environment. For example, it is estimated that around 90% of the volume used for cooling thermal and nuclear power plants is returned to the watercourse, while irrigation returns a small portion of its abstracted volume to the environment. The latter explains why rates for cooling purposes are quite low compared to other uses, but does not explain why rates for irrigation are lower than rates applied to drinking water supply, knowing that almost all volumes used by households are returned to the environment after treatment.

Users from the industrial and agricultural sectors are commonly overrepresented in RBCs compared to households, which may explain higher rates applied to drinking water supply.

Ceiling (maximum rates) are set by the national law. It could be appropriate to also set minimum rates to avoid some uses to be charged close to zero.

Exemptions could be better justified. It is unclear why activities such as aquaculture, which use a lot of water, are exempted. It is not known whether impact assessments were performed to justify these decisions.

Annex 3.A2

Water abstraction charges in practice – The United Kingdom

Currently the United Kingdom (UK) has a supply-demand water surplus. Even during periods of low water flow the UK would have enough water for human uses in all except one of its catchment areas. However, in addition to the occasional drought, local variation exists and if abstraction was left unrestricted abstraction would exceed availability during low flows in several areas. However, changing rainfall patterns due to global warming and an expected population increase in the UK will increase the pressure on available water resources.[18] Figure 3.A2.1 shows the distribution of abstracted water and how it is being used for various purposes.

Figure 3.A2.1. **Distribution of abstracted water in the United Kingdom**

Estimated actual abstraction from all surface and groundwater sources in 2012 – expressed by use in Ml/d and as % of global

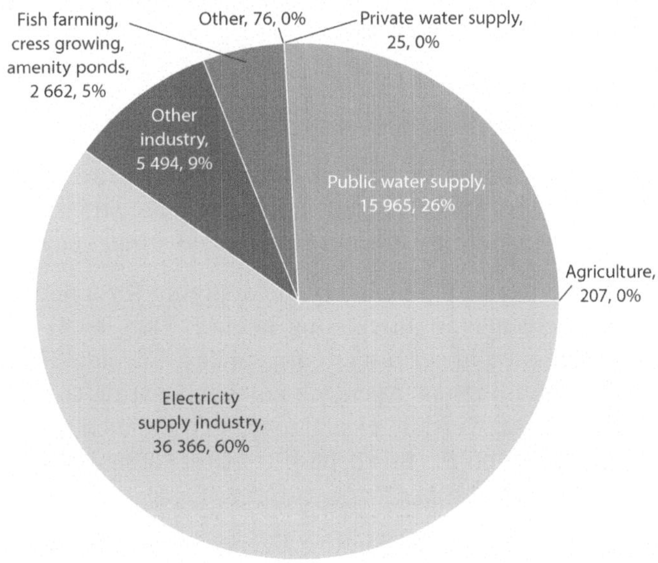

Source: Water Resources Allocation – Sharing Risks and Opportunities: Country profile UK.

This annex describes the policies and charging scheme in use to regulate water abstraction in England and Wales, which have a distinct abstraction system; Scotland is regulated separately. Responsible authorities are the Environmental Agency and Natural Resources Wales.

Regulation of water abstraction

The UK regulation of water abstraction is designed to make sure that abstraction is sustainable and does not damage the environment. Managing water abstraction is done through a licensing system, and central to the planning and management of these licences are river basin management plans (RBPMs) and catchment abstraction management strategies (CAMS). A river basin district (RBD) covers an entire river system, which in addition to the river include lake, groundwater, estuarine, and water bodies. England and Wales currently have 11 RBDs (revised every 6 years) and the RBMPs are designed to protect and improve the quality of the water environment.[19] CAMS are used to assess the amount of water available for further abstraction licensing. Hence, the CAMSs are essential to addressing both environmental and scarcity issues through the abstraction licences.[20]

The Environmental Agency assesses applications for abstraction against local water availability.[21] Licences are time-bound, with most being of 12 years duration, but they can vary, usually between 6 and 18 years. Some abstraction is exempted from licence:[22] abstraction of 20 cubic metres or less a day (provided the abstraction is part of a single operation, or if multiple points the combined total); some land drainage operations (e.g. flood protection); filling ships or boats with drinking or ballast water; water used for firefighting; abstraction in relation to dewatering quarries, mines and other building or engineering operations; and trickle irrigation (when considered "relevant projects" for the purpose of water resources).

A licence defines the maximum amount of water that may be taken in a given time period; the maximum may vary according to the time of year. Most licences require that abstraction be metered and that records are kept to monitor compliance. Sanctions for non-compliance include both criminal sanctions, like warnings and formal cautions and prosecutions with fines, and civil sanctions.

Abstraction charges

The levels of the charges for the abstraction licence holders are set to recover the costs for managing water resources. Three charges apply to the abstraction licence process:[23] application charge; advertising administration charge; and annual charge.

The application charge is payable by everyone applying for a new licence to abstract (or impound) water or to changes in an existing licence. Two levels of the charges exist and the charge can be exempted in certain circumstances. The advertising administrative charge should cover the Environment Agency's costs in relation to its costs to advertise applications for licences. These two charges are not directly related to the quantity of water abstracted. This report focuses on the third charge, i.e. the annual charge, which is related to the volume of water abstracted.[24]

The annual charge is payable in relation to a licence, expect for water authorised to be abstracted for electricity production of not more than five megawatts, for inland waters having an average chloride content in excess of 8 000 milligrammes per litre, a temporary licence, or a transfer licence. The annual charge is the sum of two elements, the standard charge and the compensation charge. Each of these elements is calculated based on a set of factors described below. The standard charge recovers the Environment Agency's costs of managing and regulating abstractions and its level should be proportional to the impact of that licence on water resources. The compensation charge adds to the standard charge, to recover compensation associated with claims of water disruptions by such businesses as fish farms, power stations.[25]

The annual charge (AC) is calculated as follows:

$$AC = \underbrace{V \times A \times B \times C \times SUC}_{\text{Standard charge}} + \underbrace{V \times B \times C \times D \times EIUC}_{\text{Compensation charge}}$$

where:[26]

- V = annual licensed volume (per 1 000 m³)
- A = source factor, classified as supported (factor 3.0), unsupported (factor 1.0), and tidal (factor 0.2)
- B = season factor, with three categories, Summer (between 1 April and 31 October inclusive, factor 1.6), Winter (between 1 November and 31 March inclusive, factor 0.16), and All Year (factor 1.0). Abstraction can be restricted to specific seasons, or if authorised both during summer and winter then charges should be calculated separately in respect of each period
- C = loss factor, relates to the purpose for which the licence is granted, with four categories, High loss (e.g. spray and trickle irrigation, and uses where evaporation prevents any return flow; factor 1.0), Medium loss (e.g. public and private water supply, commercial purposes, bottling and uses which incorporates water in the product, etc.; factor 0.6), Low loss (e.g. mineral washing, vegetable washing, and non-evaporate cooling; factor 0.03), and Very low loss (e.g. power generating of greater than 5 megawatts, fish farms, water meadows, etc.; factor 0.003)
- D = adjusted source factor, non-tidal (supported/unsupported, factor 1.0), tidal (factor 0.2)
- SUC = standard unit charge (GBP/1 000 m³)
- EIUC = environmental improvement unit charge (GBP/1 000 m³)

The Standard Unit Charge (SUC) and the Environmental Improvement Unit Charge (EIUC) are regional and shown in Table 3.A2.1 below. They can be revised every financial year.

Table 3.A2.1. **2016-17 Water charges in England and Wales**

GBP/1 000 m³

Regional charging area	Standard Unit Charge	Environmental Improvement Unit Charge
Anglian	27.51	13.71*
Midlands	14.95	0
Northumbria	29.64	0
North West	12.57	3.86
Southern	19.23	12.11*
South West (incl. Wessex)	19.71	12.91*
Thames	13.84	0.83
Yorkshire	11.63	0
Dee	13.58	0
Wye	13.58	0

* The EIUC charges for non-water undertakers in the Anglian, Southern and South West charging areas will be held in abeyance and not be levied pending further changes to the Abstraction Charges Scheme.

Source: Abstraction Charges Scheme 2016-17.

The Environment Agency provides an Abstraction charges calculator. The calculator is available online[27] and Figure 3.A2.2 provides an example of an annual charge where the volume of the licence has been specified. By varying the different components of the calculator the charges will vary. If a licence authorises abstraction with various charge factors, then the total charge will be calculated from the separate charges. However, if the licence does not specify separate annual quantities then (unless authorised by the Environment Agency) the highest charge factor will apply. The minimum annual charge (as of 1 April 2016) is GBP 25.

Figure 3.A2.2. **Charge calculator**

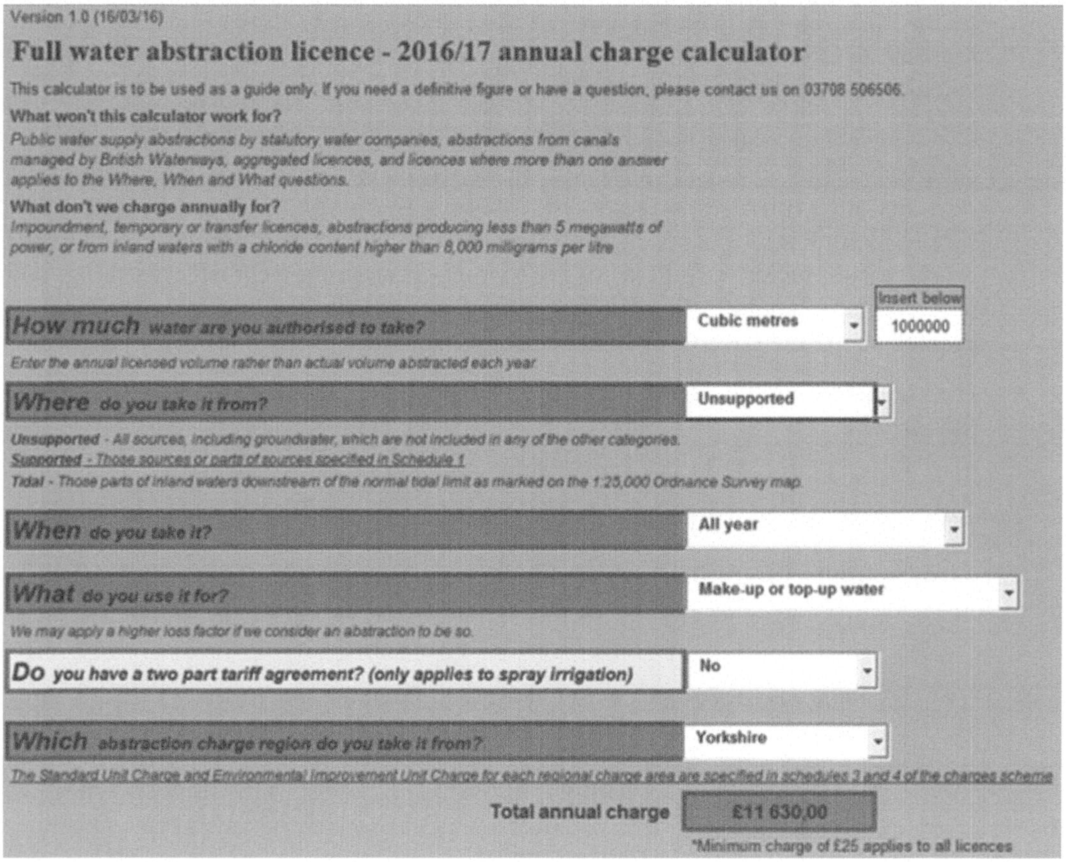

Source: Abstraction Charges Scheme 2016-17.

Strengths and weaknesses of the UK system and its future

The UK water abstraction licensing system is designed to comply with the European Water Framework Directive. Water abstraction's impact on the environment is taken into account and charging scheme is based on the "Polluter Pays principle". The estimation of the annual charge allows for a diversification of the licence charge related both to the environmental impact of the water abstraction and its effect on water scarcity.

However, the objective of the pricing structure is to recover the cost of water management. That is, the objective is not for the charges to reflect the true social cost of water abstraction, including scarcity, environmental impacts, and other externalities. Water scarcity is addressed by changing the amount of water available and by the option to revise the factors of the annual charge calculator. This "command and control regulation" is complemented by the possibility to trade licences.[28] Trading is encouraged since it allows for a more efficient allocation of water resources without additional abstraction.

The environmental component of the charge, i.e. the EIUC, may be a poor indicator of true environmental costs, and as shown in Table 3.A2.1 it is even estimated to be equal to 0 in several regions. The EIUC is based on compensation claims made by firms, and hence, will not capture environmental cost not related to financial costs to firms. Moreover, they are likely to not capture recreational values (swimming, kayaking, etc.). Hence, the EIUC may significantly underestimate the true social environmental cost and hence not fully comply with the "Polluter Pays principle".

The UK water abstraction licence system is currently undergoing major revisions. The new system is expected to be implemented in 2018.[29] However, it is expected that the seasonal permit systems will be abolished; instead water will be allowed to be abstracted when flows are high. The EIUC will also disappear, and the trading system will be reformed to allow for quicker and easier trading.[30]

Notes

1. Irrigators are supposed to pay but have been exempted (see below).

2. In consideration of farmers' financial burden, agricultural users have been exempted from water charges by Presidential decree. The cost unrecovered due to exemption is mostly supported by public budget. OECD (2010) notes: "In Korea, the 60% of rice lands in larger schemes under the management of the Korea Rural Community Corporation (KRC) have been exempted from agricultural water charges (excluding mandatory labour levies) since 2000. The smaller schemes outside KRC domain, managed under the oversight of local governments by Irrigation Associations or by individual farmers, remain largely responsible for covering all their capital, O&M and labour cost." MAFRA notes that, i) based on public consensus, farmers have transferred the irrigation association's assets to KRC, and revenues from the assets cover 50-55% of the O&M costs of irrigation facilities every year; and ii) after 2000, farmers in the areas managed by KRC stopped paying water tariffs directly, yet they still indirectly pay water tariffs.

3. Tariffs for dam water and multi-regional water systems are considered public utilities charges. 11 public utility charges are subject to the Price Stabilization Act: Electricity, Gas (bulk), Railway, Intercity bus, Express bus, Highway toll, International airline, Water (dam water, multi-regional water), Post, Telecommunication, Broadcasting.

4. The Korean consumer price increased by 27.5% over the same period.

5. A water bill examined by the authors displays tariffs for water supply, wastewater, water use charge, and eventual discount. Dam water and multi-regional water tariffs do not feature on the bill.

6. According to the interviews organised in the course of the Policy Dialogue, discussions in the water charge review committee are driven by government departments. It is difficult for other representatives to express their opinions, particularly because they receive limited advance information on time to prepare for meetings.

7. To clarify, user 1's consumption increases from left to right, whereas user 2's consumption increases from right to left. Hence, both users' demand curves are downward sloping.

8. The description of the non-market valuation approach and its techniques is brief and highly simplified. For details about the approach and the different methods presented here, see, e.g. Freeman et al. (2014).

9. Again, the description here is highly simplified. Care would need to be taken to make sure that everything else is equal, and other benefits and the costs of the programme not reflected in the income differences should be considered.

10. The abstraction charges of public projects (national projects, publicly-owned projects, and irrigation projects) are usually free or reduced (OECD, 2015b).

11. The Netherlands impose a tap water tax (EUR 333/1 000 m^3 in 2015) which is levied from water suppliers for up to 300 m^3 a year per connection to the water system. Large users are basically exempt from the tax. Provinces also charge a groundwater levy for large groundwater abstractions, which applies mostly to industry and drinking water companies. The agricultural sector is not charged for groundwater abstraction.

12. The total amount raised from the central government's groundwater tax was about EUR 180 million per year.

13. The first priority use relates to ensuring safety and preventing irreversible damage such as ensuring the stability of flood defence structure and the settling and subsidence of peat bogs and moorland (OECD, 2014b).

14. A national water charge would make sense if the water network is so well interconnected that it is possible to transfer (easily and at a low cost) some water from one point to another. In that case, regionalisation is not needed since the public authority manages a single resource. The abstraction charge simply reflects the global scarcity of water for the whole system. Under other circumstances, as illustrated in Box 3.6, a uniform tax will not achieve economic efficiency.

15. At a later stage, based on information on differentiated costs of the resource, Korean authorities may initiate a dialogue about the opportunity of varying water abstraction charges across hydrographical zones. Within a given hydrographical zone, water abstraction charges will vary depending upon zones of specific interest from a water management point of view, e.g. as the ZREs in France.

16. Based on OECD (2015b).

17. Based on Ministère de l'Ecologie, du Développement Durable et de l'Energie (2012); Cour des Comptes (2015).

18. For more information, see www.ofwat.gov.uk/wp-content/uploads/2015/11/pap_pos20111205abstraction.pdf.

19. https://www.gov.uk/government/collections/river-basin-management-plans-2015.

20. For more detailed information on the CAMS approach and how areas are classified, see https://www.gov.uk/government/uploads/system/uploads/attachment_data/file/297309/LIT_4892_20f775.pdf.

21. For more details on the management of water abstraction, see reference in the previous footnote.

22. https://www.gov.uk/guidance/water-management-abstract-or-impound-water.

23. For details see: https://www.gov.uk/government/uploads/system/uploads/attachment_data/file/509928/LIT_9909.pdf

24. As of 1 April 2016 the lower and higher application charge is GBP 135 and GBP 1 500 respectively, and the advertising administration charge GBP 100.

25. Personal communication with representative of the Environment Agency (2016-06-03).

26. See Abstraction Charges Scheme 2016/17 for details https://www.gov.uk/government/uploads/system/uploads/attachment_data/file/509928/LIT_9909.pdf.

27. https://www.gov.uk/government/publications/abstraction-charges-scheme-april-2014-to-march-2015.

28. http://webarchive.nationalarchives.gov.uk/20140328084622/http://cdn.environment-agency.gov.uk/geho0711btzk-e-e.pdf

29. Personal communication with representative of the Environment Agency (2016-06-03).

30. https://www.gov.uk/government/uploads/system/uploads/attachment_data/file/492411/abstraction-reform-govt-response.pdf.

References

Cour des Comptes (French Court of Auditing) (2015), *Les agences de l'eau et la politique de l'eau : une cohérence à retrouver.* Rapport public annuel, Paris.

Chae, S.B. and Kang, K.R. (2011), The Estimation of the Economic Value of Avoiding Cutting off the Water Using Contingent Valuation Method (조건부 가치추정법을 이용한 단수상황 회피의 경제적 가치 추정 연구), *Seoul Studies*, 12(2): pp. 141-153.

European Environment Agency (2013), *Results and lessons from implementing the Water Assets Accounts in the EEA area*, EEA Technical Report no. 7/2013, Publications Office of the European Union, Luxembourg.

Freeman, A.M., Herriges, J.A., and Kling, C.L. (2014), The Measurement of Environmental and Resource Values, *RFF Press*, New York, NY, USA.

International Office of Water (2009), *Organization of Water Management in France*, Paris.

Kim, H-R (2013), State-centric to Contested Social Governance in Korea: Shifting Power, *Routledge Studies on Modern Korea*, Routledge, New York.

Kim, S. (2010), Collaborative Governance in South Korea: Citizen Participation in Policy Making and Welfare Service Provision, *Asian Perspective*, Vol. 34, No. 3, 2010, pp. 165-190.

K-water (2015), *Water for a Happier World. K-water Sustainability Report 2015*, K-water.

Kim, B.K., Cho, Y.S., and Kwak, J.E. (2001), Estimation of WTP for Water Quality Improvement in Paldang Reservoir Using Contingent Valuation (팔당호 수질개선에 대한 소비자 지불의사액 추정), *Environmental and Resource Economics Review*, 10(3): pp. 433-459.

Kwak, S.J. (1994), Contingent Valuation of Improving Safety of Tap Water Quality in Seoul (조건부가치측정방법을 통한 수질개선에 관한 편익추정 – 서울지역의 수돗물 안전성증가에 대한 비용편익분석을 중심으로), *Korea Spatial Planning Review*, 21: pp. 23-40.

Lee, S.C. (2009), *Water Use Charge System and Basin Management in Korea – Comparative Analysis with the Forest and Water Source Tax in Japan*, Research Center for Sustainability and Environment, Shiga University.

Lee, Y., Hong, Y.S., and Chung, Y.K. (2015), The Economic Value of an Ecological River Restoration Project: Applying the Individual Travel Cost Method (안양천 복원사업의 경제적 가치추정: 개별여행비용법을 중심으로), *Journal of Environmental Policy and Administration*, 23(4): pp. 97-114.

Ministère de l'Ecologie, du Développement Durable et de l'Energie French Ministry of Ecology, Sustainable Development and Energy (2012), *Tax on water abstraction : how can it be used in the quantity management of the resource?* Department of the Commissioner General for Sustainable Development no. 127, Paris. Available at www.developpement-durable.gouv.fr/IMG/pdf/LPS_127_redevanceEVcorrige-AB.pdf (accessed 14 September 2016).

OECD (2010), *Agricultural Water Pricing: Japan and Korea*, OECD Publishing, Paris. http://dx.doi.org/10.1787/9789264083578-13-en.

OECD (2015a), *Stakeholder Engagement for Inclusive Water Governance*, OECD Studies on Water, OECD Publishing, Paris. http://dx.doi.org/10.1787/9789264231122-en.

OECD (2015b), *Water Resources Allocation: Sharing Risks and Opportunities*, OECD Studies on Water, OECD Publishing, Paris. http://dx.doi.org/10.1787/9789264229631-en. Country profile (France) available at: www.oecd.org/fr/publications/water-resources-allocation-9789264229631-en.htm (accessed on 14 September 2016).

OECD (2014b), *Water Governance in the Netherlands: Fit for the future?*, OECD Studies on Water, OECD Publishing, Paris. http://dx.doi.org/10.1787/9789264102637-en.

OECD (2012), *Meeting the Water Reform Challenge,* OECD Studies on Water, OECD Publishing. http://dx.doi.org/10.1787/9789264170001-en.

OECD (2012), *A Framework for Financing Water Resources Management, OECD Studies on Water*, OECD Publishing. http://dx.doi.org/10.1787/9789264179820-en.

OECD (2011), *Water Governance in OECD Countries: A Multi-level Approach*, OECD Studies on Water, OECD Publishing. http://dx.doi.org/10.1787/9789264119284-en.

Ofwat (2011), *Bulk supply pricing – a statement of our policy principles*, Ofwat, Birmingham.

Park, D.H. and Park, Y.S. (2007), Willingness to Pay for Residential Water on Drought Conditions (가뭄시 가정용수에 대한 소비자 지불의사), *Journal of Korea Water Resources Association*, 40(11): pp. 861-867.

Pyo, H.D. and Choo, J.W. (2014), Estimating Willingness-to-Pay for the tap water quality improvement in Busan using contingent valuation method (조건부가치측정법을 이용한 부산시 상수도 수질개선에 대한 WTP 추정), *Journal of Korean Society of Water and Wastewater*, 28(5): pp. 561-571.

Pyo, H.D. (2015), Estimating the Economic Value of the East Sea Coast by Using Travel Cost Method (여행비용법을 이용한 동해안 해수욕장의 경제적 가치추정), *Journal of the Korean Society for Marine Environment and Energy* (11): pp. 65-71.

Rogers, P., R. Bhatia, A. Huber (1998), *Water as a Social and Economic Good: How to put the principle into practice,* TAC Background Papers no. 2, Global Water Partnership/ SIDA, Stockholm.

Souza d'Alte T. (2010), *Report on Portuguese Environmental Economic Instruments*, Agencia Portuguesa do Ambiente.

Water Account Australia (2016), *Water Account, Australia, 2014-15*, www.abs.gov.au/ausstats/abs@.nsf/mf/4610.0.

Chapter 4

Smart Water Management in Korea

This chapter presents Smart Water Management in Korea, an initiative to combine water and information and communication technologies to improve water resources management and the supply of water services in Korea and abroad.

The chapter identifies some bottlenecks, which have to be overcome to expedite the deployment of smart water management in Korea and abroad.

1. The Smart Water Management in Korea: State of play

The Smart Water Management (SWM) in Korea seeks to establish a new paradigm for water resources management, in a context marked by more uncertainty about future water availability and demand as well as increased risks of scarcity and floods generated by climate change. In Korea, the context is also marked by limited options to augment water supply in the face of growing water demand. Therefore, there is a strong case for SWM, which echoes the issues addressed in Chapter 3 on how to make the best use of economic instruments to manage water quantity in Korea.

That concept is valid domestically and abroad. As noted in the *OECD Environmental Outlook to 2050* (OECD, 2012), competition to access water resources intensifies in most developed and developing countries. Climate change adds to the complexity and calls for adaptation measures that make the best use of available water and infrastructure, and adjust to shifting circumstances at the least cost for society. Therefore, SWM has a potential to be exported in countries facing similar challenges (see Box 4.1).

Box 4.1. Market trends for smart water systems

The smart water market as a whole can be seen as fragmented both in terms of the number of players and their market share and in the lack of cross-sectoral applications (i.e. infrastructure monitoring, river monitoring, domestic water metering, industrial and commercial consumers, agriculture, recreation and gardening). While companies are all based upon the principle of acquiring, analysing and distributing data, companies to date have tended to concentrate on specific applications.

At the venture capital level, the water sector's share of CleanTech funding has been between 2.0% and 3.4% in recent years. Within water investments in 2010, 47% went on water treatment (USD 121 million), 41% on wastewater treatment (USD 105 million) and 12% on smart water and smart infrastructure (USD 31 million). Between 2006 to 2010 "smart technology" accounted for 11% of total CleanTech venture capital funding (USD 3 910 million out of USD 35 210 million) with 2% of this going on Smart Water (USD 80 million).

It is estimated that global sales for smart water systems were in the range of USD 500-1 000 million in 2009-10 and they are forecast to rise to USD 5-16 billion by 2020. Smart water systems accounted for approximately 0.5-0.9% of the global water hardware market in 2010 and look set to account for 2.9-9.4% of the market by 2020.

Smart networks and the associated management of large quantities of data open business opportunities for the water industry and new comers. For instance, energy or telecommunication companies increasingly see themselves as actively managing devices in the home; and smart homes could be managed as integrated wholes covering water, energy and other services. It is not clear how market power and influence will change, with the deployment of smart water systems.

Source: OECD (2012), *Policies to Support Smart Water Systems. Lessons from Countries Experience*, unpublished paper by David Lloyd Owen commissioned by the OECD.

SWM is designed to tackle increased uncertainties and risks of water-related disasters by developing systematic and effective response mechanisms in a sustainable manner. It does so by making the best use of information and communication technology (ICT) to produce and use large volumes of data in real time to support (and integrate as much as possible) water resources management at different scales, from dam management and flood prevention to detection of leakages and promotion of water use efficiency in homes. When fully deployed in Korea, it is expected that SWM will be able to minimise or postpone the construction of additional dams and flood defences, by making the best use of available assets and water resources. SWM is well aligned with Korea's national policy towards "a creative economy", which endeavours to create value and jobs through innovation that combines information and communication technology.

In practice, SWM is being used as means to achieve effective goals with Integrated Water Resources Management (IWRM) for future water issue. K-water is developing a distinctive approach to SWM in Korea, hereafter referred to as K-SWM.

On the one hand, dam integrated management – an element of IWRM – combines water infrastructure and information and communication technology to effectively manage risks of floods and droughts, by anticipating water demand and availability in real time. Water security in Korea, in particular the management of flood and scarcity risks, heavily depends on an extensive system of infrastructure. The capacity to operate and manage that system effectively determines Korea's resilience to water risks. ICT can enhance that capacity by collecting and sharing real-time data on water use, expected rainfalls and available room in reservoirs. That information is used to manage dams and water bodies, minimising risks of floods and droughts. This application enhances K-water's capacity to deliver its mandate on water quantity management and prevention of scarcity and flood risks. It can serve similar purposes abroad, where other agencies have a similar mandate.

On the other hand, the smart water city project – the representative example of K-SWM – provides final water users with instant and reliable information on water use and the quantity and quality of water supplied, in particular to facilitate leakage detection and encourage tap water drinking. Water suppliers in Korea are confronted with the need to renew existing infrastructure, as ageing affects its performance, in particular as regards leakage. As renewal needs to take place in a context of fiscal consolidation that makes public finance scarcer than before, revenues from water tariffs will have to play an important role in financing the renewal and upgrade of water supply and sanitation in the coming years/decades. One way to increase revenues while keeping prices low is to minimise non-revenue water (leakage) and to increase consumption (in particular for drinking purposes). ICT can help detect leakage and inform (domestic) water users about water use and water quality. Although SWM is now led by K-water, it can operate in a competitive environment in the future, where other suppliers of water or communication services may be able to develop similar applications when there is a demand from cities and households.

K-SWM was first rolled out in Korea in 2015. A situation room at the Water Resources Operations Centre manages dams operated by K-water nationwide: the Centre forecasts rainfall in the basins of 58 dams and weirs nationwide, gathers and manages data from more than 500 hydrometric stations as well as controls and manages floods, water supply and hydropower generation of multipurpose dams, water supply dams and weirs. In addition, the Centre develops technologies of integrated water management based on ICT and exports these technologies abroad. Several cities are pilot-testing applications for domestic water users, including smart water meters that monitor water quality in real-time. Users see some value in improved services that increase trust in the system and willingness to drink tap water. However, at the initial stage, the transaction costs are significant (the time spent by K-water staff to promote and roll out K-SWM in pilot cities). The financial costs matter as well, as investment in smart water meters is covered by public funds. As K-SWM is rolled out beyond pilot cities, the business model should allow recouping these costs by efficiency gains or additional revenues.

It should be noted that, as for any environment-related innovation, environmental performance is best rewarded when the policy framework reflects the environmental externality (the cost for the community of pressures on the environment). Therefore, water charges that fully cover the costs of supply (including the opportunity costs) can only make K-SWM more attractive for users.

2. Policies to support eco-innovation. Relevance for Smart Water Management in Korea

The section draws lessons from the experience of OECD countries with policies to support water-related innovation, which are relevant for SWM. The OECD (2011) notes that market mechanisms alone will not provide an appropriate amount of eco-innovation at the right time. This is because innovators may not reap all the benefits of their innovations, and because environmental benefits may not be appropriately valued by markets. This is particularly the case for water-related innovation, where the opportunity costs and environmental costs of using or polluting water are not reflected in prices paid by water users. Since markets fail to deliver the appropriate level of environment-related innovation, policy interventions are required. The question then is: what is the best way to support the development and diffusion of eco-innovation?

The OECD has reviewed experience of several member countries (Australia, Canada, Japan, Korea, Mexico, New Zealand, Turkey and the US) with policies to support eco-innovation.[1] Some generic features come out of these reviews, which can inform policy discussions in Korea about the development and deployment of SWM.

The rationale for policies to support eco-innovation

OECD countries that actively support the development and the diffusion of environment-related innovation do it for two purposes. First, eco-innovation can help address domestic environmental challenges; they contribute to environment policy objectives. Second, eco-innovation can create new and sustainable jobs; after 2008, it has been considered a new engine for growth, as crystallised in the concept of green growth. The potential for growth is all the more important when markets for a particular innovation are global and growing. Relatively small countries, with limited domestic markets for eco-innovation, are targeting global markets and set out policies to grow their global market share. This is typically the case for Denmark for wind energy technology.

This line of reasoning supports policies for water-related innovation. Research commissioned by the OECD has measured the costs of too much, too little, or too polluted water, or the costs associated with lack of access to water supply and sanitation services (Sadoff et al., 2015). This is a global issue, as developing and developed countries are affected. The costs of water insecurity are particularly high in fast growing Asia. China and India top the list of countries facing the largest costs of water insecurity. Therefore, as smart water management can help minimise these costs, it potentially faces a global demand. That demand will materialise when the needs for smart water management can be monetised. This is where the market failures signalled above have to be overcome, and public policies are required.

It follows that Korea may benefit from investing in smart water management: Smart Water Management can address water management issues, now and in the future, minimising the costs associated with water risks and investments in water security, and letting the benefits of water security materialise; smart water management can grow as an export-driven industry, to address demand in Asia and globally.

Selected features of policies to support eco-innovation

The OECD notes that eco-innovation policies have to adjust to the features of the domestic economy, in particular the knowledge base, the size of domestic markets, and the vigour of the venture capital industry. From that perspective, the water industry landscape in Korea is contrasted. On the one hand, the knowledge base and technology frontier are well advanced, compared to most OECD countries. Korea is a leading country when it comes to ICT. It has also been able to develop and export water-related goods. The domestic market is limited by the size of the country and population. However, rapid economic growth and urbanisation have stimulated households and business capacities to invest in environmental goods and services. On the other hand, private capital and entrepreneurs have not been involved in the development and roll-out of SWM so far. These features drive the way SWM can develop.

In most non-EU OECD countries, public research and development (R&D) remains a major orientation of policies to support environment-related innovation. The US and Japan typically allocate significant public finance to environment-related R&D. However, three trends have emerged: i) some countries are concerned by the competition and trade issues related to such support; ii) public resources are increasingly channelled via departments not directly in charge of environment policies (Energy, Agriculture, Transport), making inter-agency co-operation a must; iii) the role of research organisations is being redefined, to intensify linkages with the private sector and stimulate the development of marketable outputs; incubators in the US, or the National Institute of Advanced Industrial Science and Technology's (AIST) Technology Licensing Office in Japan illustrate innovative arrangements in this area. *SWM fits well into this context, as it combines public research and innovation from the ICT industry. It stimulates the development of an operational tool, which is close to market and diffusion.*

Attracting private funds to finance environmental R&D is another major policy orientation. The main issue is to reduce risks for private investors investing in environmental R&D projects, while making sure that public money is used effectively and does not crowd out private initiatives. A variety of funds were established to reduce risks to private investors (e.g. Sustainable Technology Development Canada-SDTC in Canada), or incubators (e.g. The Clean Energy Alliance in the US, Environmental Technology Business Incubator in Korea). Measures are taken to stimulate the venture capital industry and to provide incentives for environment-related projects; e.g. this is the role of the Environmental Venture Fund in Korea. *So far, K-SWM has not explored measures for private funding and relies heavily on K-water's financial resources.*

Environment-related performance standards are being set with the aim of stimulating innovation in the goods and services sectors. Such standards are pursued in such fields as energy and resource efficiency. *Standards related to the quality of water supply services and non-revenue water can stimulate more systematic roll out of SWM and related technologies to detect leakage and respond to consumers' demand.*

In line with the Recommendation of the OECD Council on Improving the Environmental Performance of Public Procurement [C(2002)3], green procurement initiatives are increasingly seen at local and national levels. Guidelines are supported by websites, green products databases, and pro forma requests for tenders. The Green Purchasing Network arguably is the most active international network in this area. *Initiatives in this area could enhance K-SMW's competitiveness vis-à-vis standard water supply services for Korean cities.*

Some initiatives set out to remove obstacles to the deployment of environment-friendly technologies and products abroad. Shared definitions, standards and labels contribute to a level playing field for the creation and diffusion of environment-friendly technologies, products and life-styles. However, such efforts are still plagued by institutional problems related to intellectual property rights and international monetary transfers. Few co-operation projects reach developing countries (with the exception of East Asia, and China in particular). Global standard development *would ultimately benefit K-SWM. It could increase the demand for similar technologies and management approaches, growing a market for smart water management.*

To conclude:

- SWM in Korea (K-SWM) fits with domestic and international needs. It has the potential to serve a global and growing market.
- Efforts made in (public) research need to be paralleled by support to create a solvable demand for K-SWM. Signalling the opportunity costs and the environmental costs of water (see Chapter 3) can contribute to creating such a market in Korea. Environmental standards and standards related to the quality of water supply and sanitation services also have a role to play in this respect.
- Public procurement in Korea can play a role, when it values environmental performance and efficient water use.

3. Policies to support smart water management in selected countries

International good practices can inform policy discussions on how to best support K-SWM. The section presents policies in several countries, which promote smart water management. It covers a range of initiatives and ambition, some at national levels and others at state or local levels. The emphasis is often on smart water management in the home, as few countries support the full spectrum of smart water management. The following section draws policy lessons.

Canada: vigorous policy to promote green technologies

Eco-innovation in Canada has improved over the past decade, with a growing clean technology sector and greater private sector investment in research and development. Revenues of Canada's clean technology industry were estimated at CAD 13.27 billion in 2015, employment of 55 200, and exports at CAD 6.7 billion (Analytica Advisors, 2017). However, clean technology companies have been hampered by a lack of domestic demand for clean innovations.

The government plans to expand support for business innovation networks and clusters, promote the clean technology sector and the adoption of clean technologies, and increase funding to support innovation and growth-oriented firms. A consultation process was launched in June 2016 to help inform development of *the Innovation Agenda* (OECD, 2016c). A specific clean technology strategy for Canada's natural resource sectors is also being developed by 2017 (ECCC, 2016). Canada's Federal Budget 2017 includes an Innovation and Skills Plan that builds on the Innovation Agenda mentioned above. A series of specific measures to support clean tech in Canada are summarised on the Government's website. In addition to these measures, Canada's Regional Development Agencies have been mandated to invest CAD 100 million per year in Clean Technology, as part of a broader contribution to the Innovation and Skills Plan at the regional level.

Ontario has established a globally-competitive water technology sector, with approximately 900 established water technology companies and 300 early-stage water innovation companies, with many selling to global markets. The provincial water industry is estimated to support approximately 22 000 jobs, and is poised to grow. The success of water companies in the province resulted from a combination of progressive water protection legislation that created demand for solutions across its 750 water and wastewater treatment facilities, a strong research and development support system, and skilled entrepreneurs (Invest Ontario, 2016). The Walkerton Clean Water Centre – established in 2004 following e.coli drinking water contamination that killed 7 and made 2 300 people ill – has played a key role in water-related training, applied research and technology demonstration.

In 2010, the province of Ontario passed the Ontario Water Opportunities and Conservation Act, aiming to co-ordinate regulation and policy with sustainable water management. This includes water sustainability plans, standardised information about water use on bills, water efficiency standards and a Water Technology Acceleration Project (WaterTAP) to support innovation by companies. The province supports venture capital (VC) funding through the Ontario Venture Capital Fund, the Ontario Emerging Technologies Fund, the Investment Accelerator Fund and the Innovation Demonstration Fund and research through 150 researchers at 25 water-related research institutes.

An additional programme for consideration is the Ontario Government's *Showcasing Water Innovation* programme, which complemented the Ontario Water Opportunities Act. The programme was intended to demonstrate leading edge, innovative and cost-effective solutions for managing drinking water, storm water and wastewater systems in Ontario communities (https://www.ontario.ca/page/showcasing-water-innovation). In 2013, the Ontario Government published a Water Sector Strategy.

This case illustrates how a state government can encourage smart water development through supporting utilities and research institutes, along with funding vehicles to create a concentration of research and funding interest in the sector. In this case, the area is water rich, but water leadership was identified as a policy objective.

Israel – supporting smart technologies

Israel is a water scarce country with recent years of drought and the potential for other threats to its water supplies. Policies to support water-related innovation in Israel combine strong pricing policy to stimulate water efficiency and demand for water-saving technologies, awareness raising, in particular in schools, and support for the development, diffusion and expert of water-related technologies.

Water scarcity has created markets for water-saving technologies for domestic and municipal uses. Economic incentives designed to reduce water demand in the urban and in the agricultural sectors based on increasing block tariffs resulted in the development of innovative water management devices, such as water meters that are read remotely and more accurately (including measuring small drops, so leakages would be fixed), pressure optimizers devices, computerised irrigation systems, etc. (OECD, 2010b).

Developing smart technologies is supported by an inter-ministerial committee. Israel's Mekorot established the Water Technologies Entrepreneurship Centre (WaTech) in 2004 as a platform for business ventures. In its first three years, 250 projects were assessed and 35 of these are being trialled by Mekorot. The utility sets technical standards at the start of the trial and their deployment is seen as a tightly disciplined and organised process in order to minimise the potential for innovation risks.

In 2006, Israel launched the Novel Efficiency Water Technologies (NEWTech) programme which has seen in 26 government-funded water technology incubators, gaining USD700 million in private investment. NEWTech will reimburse 70% of installation costs (up to USD 200 000) in order to mitigate the risk in installing innovative technologies. While it predates NEWTech, Israel's main incubator is Kinrot, one of the most active venture firms globally. Not only has Kinrot established partnerships with global equipment vendors like General Electric, it boasts over 20 companies in its portfolio. Companies such as TaKaDu (leakage detection and management) and Whitewater (real time warning systems) have emerged from this policy support.

Other water policies also triggered private sector response and illustrate the benefit of economic instruments (fines for water leakage, or tariffs that reflect scarcity):

- Water loss fines for municipalities at a level of above 12% water loss created incentives for development of water loss detection and dynamic water pressure equipment.
- Several consecutive years of drought led to a significant increase in water prices. In 2009, an additional "surplus use" fee has been imposed on domestic uses, to discourage excessive water consumption. During these years, one could observe establishment of many water technology start-ups and also implementation of technologies at all scales – from home water-saving devises to accurate reading of water meters to establishment of new desalination plants.

Malta – an integrated water and electricity smart grid

Malta is a water-short island with a population of 400 000 with utility services for 245 000 electricity and water customer accounts. The country is dependent on imported oil for its energy and desalination along with increasingly depleted groundwater resources for its water. The groundwater is also suffering from saline ingress and its viability is set to be increasingly compromised from 2020-25. Groundwater accounts for 60% of supplies, desalination (reverse osmosis) for 30-35% and surface water via cisterns, 5-10%. As electricity accounts for 75% of the reverse osmosis plant operating costs, when electricity rates for large customers were raised by 60% this resulted in a significant increase in overall water operating costs and water bills rose by up to 25%.[2]

A Climate Change Committee was appointed by the Ministry for Resources and Rural Affairs in 2008. A report on measures required was presented to the Minister in 2008 and in 2009 consultation process was undertaken resulting in a national smart utility metering plan, to be rolled-out between 2010 and 2013. This is understood to be the first comprehensive policy case at the national level regarding smart water policy and the world's first universal smart water plan.[3]

The smart metering project was part of a package of 96 measures including four specific demand management policies for water (mandating rainwater capture, fiscal incentives for using well water in domestic cleaning and flushing and various measures to encourage efficiency and reduce water demand). The project will cost an estimated EUR 40 million or EUR 163 per electricity and water meter. Between 2010 and 2013, all water meters will be retrofitted with digital data modules, which will send water usage data to their local zones for collection and analysis. After a trial installation of 5 000 smart meters in February 2010, 84 000 meters will be installed each year.

Smart metering is being managed by the Ministry for Infrastructure, Transport and Communications. Electricity is provided by Enemalta Corp and Water by the Water

Services Corp. Electricity and water meter reading and billing has been outsourced to Automated Revenue Management Services Ltd, a company jointly owned by the two utilities. Planning for the programme started between the utilities and IBM Global Services in 2008.

The two networks are separate but they share data handling and costs were minimised by the work being carried out jointly. Water data flow as indicated below:

Water meter → Radio transmitter → Electric meter → Electric utility → Water utility

Malta's compact nature and the transparency of the challenges facing its utilities have enabled a broad public support for these measures.

The Netherlands – pragmatism and politics

In the Netherlands smart metering systems for power grids and natural gas will be implemented, according to European law. Trial installations started in 2012 with smart meters being offered to all households from 2014 on a voluntary basis.

A co-ordinated inter-ministerial national structure places water management high on the agenda of several ministries. The Dutch public and private sectors have established the Netherlands Water Partnership (NWP). The principal aim of the NWP is to co-ordinate Dutch activities abroad and to promote Dutch expertise in the water field worldwide.[4] The Netherlands is seeking to be seen as a world leader in sustainable water management.

Policy is vulnerable to public concern

In 2010, the Netherlands had to put its universal smart water metering policy on ice because of media and "vocal minority" opposition, relating to concerns about health and privacy. As customer interests are not normally taken into account, utilities can encounter unexpected opposition and not be able to respond to it, therefore rendering policy impotent.

Smart metering installation is affected by stranded assets

Meanwhile, the municipally owned water utilities are divided about smart water systems. Some companies like Vitens, WML and PWN see opportunities in the drinking water sector. They install a smart water meter for free and see this as an opportunity to serve customers better. Amsterdam's Waternet will not install these meters since the company only recently started installing new water meters and replacing them with smart ones is seen as a waste of money.[5] This is an example of the challenge of stranded assets, where installing a new technology involves replacing hardware that is fully operational.

Singapore – smart management as a part of holistic water management

Because of the Singapore's dependence on water imported from Malaysia and its aim to be water self-sufficient by 2060, water management is regarded as a priority by the government. The state-owned Public Utilities Board (PUB) is responsible for all aspects of water management and service provision and is administered by the ministry of Environment and Water. PUB is mandated to work with companies and research institutions for developing new technologies that can assist in water management and create opportunities for Singapore based enterprises.

Using the utility as a smart technology testing platform

The PUB acts as a test bed for new technologies and their application via public private partnerships with the Economic Development Board being a facilitator and intellectual property remaining in the private sector. Amongst the 294 R&D projects carried out between 2002 and 2010, smart water projects have looked at real-time water quality monitoring and analysis, membrane integrity sensors for wastewater recovery and microbial source tracking.[6]

A smart grid to optimise water management

In order to attain water self-sufficiency in an affordable manner, all aspects of the water cycle are managed including desalination, wastewater recovery, catchment management and rainwater sewerage recovery. This also requires extensive monitoring and both management and monitoring have been integrated into a smart water grid. Policy support has come from the 2009 Active, Beautiful, Clean Waters Programme (inland water quality) and a series of water efficiency initiatives including the 2011 Water Conservation Awareness Programme. Due to the efficiency of the extent water network (distribution losses are less than 5%) and the other monitoring programmes, household smart water metering has not to date been seen as a priority.[7]

Policy development in Singapore is aimed towards long term water self-sufficiency. Adoption of smart water measures is pragmatic, being focussed towards areas where it is seen to meet their needs. Where this is the case, policy is supportive especially in assisting the development of appropriate approaches.

The United Kingdom – a case of policy grid lock?

Regulatory incoherence slows down planning for smart metering

Water and wastewater services in England and Wales were privatised in 1989. Ofwat, the sector's economic regulator, has statutory powers to set price and performance limits, which in effect determine each company's spending priorities. The sector operates within a series of five-year Asset Management Plans (AMPs), AMP5 runs from 2010 to 2015 and AMP6 will run from 2015 to 2020. In 2010 Ofwat stated that they don't see the benefits of smart metering' and therefore policy development is concerned with AMP6.

In England and Wales, a significant excess of demand over renewable supplies is forecast from 2030. The government has set a target of reducing per capita consumption from 150 litres per capita per day to 130 and acknowledges the role of metering in achieving this aim. The most pertinent example of this policy is the legislative changes which have allowed water companies to enforce mandatory metering in areas designated as "water scarce".

There are no policies that in fact impede plans to develop smart water network at the macro-legislative level. Policy promotes the general principles and drivers for smart grid. The challenge is operating within the 5-year AMP cycles, which result in five-year cyclical spending patterns. Instead of innovation, utilities have been driven to large framework agreements with a small group of outsourcing specialists, driving down margins through economies of scale.[8] Ofwat regards efficiencies as being deliverable within each AMP period after which each utility's cost base is re-set, cancelling out the efficiencies attained during the period. Smart metering has too long a payback time for this model. In addition, the ability for companies to get a return on their installed asset base can be seen to favour the development of new assets rather than the optimal management of a minimal asset base.

Since 2011, DEFRA (the environmental regulator) has acknowledged the importance and benefits of smart water metering but considers that it is premature, economically and financially, to commit to such schemes.[9] Even when meters are installed by a utility, volumetric tariffs can only be charged when the customer opts for them.

In 2011, Ofwat concluded that the greatest net benefits would be delivered from an accelerated programme moving from 38% metering in 2010-11 to 90% metering by 2029-30.[10] The review noted the potential for smart metering but did not specifically consider it. Ofwat has commissioned UK WIR (the water utility joint research body) to prepare an evaluation of the costs and benefits of smart metering before the price setting for AMP6 in 2014-15.

Southern Water – a water scarce policy exception

DEFRA supported Southern Water's 25-year Water Resources Management Plan, which is based on prioritising smart metering. In 2010, 40% of domestic accounts were metered, which is set to rise to 92% by 2015, an additional 487 000 accounts. Meters are to be installed at the property boundary and will be deployed for leakage detection.[11]

A customer awareness programme takes place before the metering programme and is being carried out by Groundwork, with the Customer Focus Creative Design Process having been developed with assistance from the Design Council. The emphasis in the consultation process is to explain the economics behind the programme. Radio and newspaper advertising is followed by distributing information and meeting local groups and communities. This is supported by a dedicated customer website and customer contact centre. It is also necessary to tailor these programmes to avoid unnecessary customer contact. Affordability issues for households facing appreciably higher tariffs have been addressed by a three year "changeover tariff" to smooth out the increase and, for households affected by real affordability challenges, there is a Support Tariff.

Scottish Water – competition as a driver

Scottish Water is a state-owned and operated company that operates within a similar regulatory framework to the privatised companies in England and Wales being regulated by the Water industry commission for Scotland (WICS). Competition for all non-domestic customers was introduced on 1 April 2008, under the Water Services etc. (Scotland) Act 2005. This affects all of the utility's 130 000 non-domestic customers. Scottish Water has announced that it will install up to 3 000 smart meters in public buildings throughout Scotland, largely based on a business case argument that water savings through better leakage control and reduced consumption will benefit both the company and its customers.[12]

The fluidity of the development of policy regarding smart water has been a major impediment to its development and deployment. There is considerable evidence of companies offering a variety of smart technologies and utilities willing to adopt them, but they have typically been held back by policy debates that have an arguable relevance to water scarcity issues affecting parts of the United Kingdom.

The United States

In the USA, national level policies have been fragmented and as of yet, there is no coherent policy around water management and conservation, let alone for smart water. Policy to support smart water management is developed and enacted at the state and local level. For example, many local water agencies and state public utility commissions (PUC) have set tiered rates designed to promote conservation. These rates typically increase the charges for water when certain threshold levels are exceeded. Most PUCs (and many localities) require that all customers are metered, although there is no national policy that dictates this. Policy at a quasi-national level is seen when the government has sought to improve sustainability within their own operations. The US military, for example, has launched a "Net Zero" programme to make facilities neutral with respect to energy and water and to reduce other waste streams.

The most significant policy objectives have been those associated with water resources management, reporting and per capita demand reduction. Policies supporting full water accounting and reporting, that require limitations on water loss, or best practices to maximise efficiency will support smart grid installation.[13]

Supporting initiatives in California and Arizona

In California, Senate Bill SB x7-7 Delta and Water Reform Legislation of 2009 mandates the state to reduce urban per capita water use by 20% by the end of 2020, along with end of 2015 interim use targets, as defined by each municipality. For the utility, it is easier to drive down consumption by eliminating data errors (missing or out-of-calibration meters, incorrect meter data, etc.), system losses and water loss through theft in order to get a true understanding of their non-revenue water, rather than undertaking wholesale pipe replacement. Under California's Proposition 218 (1996), a water rate increase must be based on the cost of providing the service. In order to demonstrate that this "proportionality" is consistent, more granular data is required, which has been seen to support the installations of smart grid technologies.

Arizona's Department of Water Resources' (ADWR) Modified Non-per Capita Conservation Program was developed between 2006 and 2008 and implemented in 2010. The programme requires water providers to adopt best management practices to achieve water conservation. While this does not specifically require advanced metering, it lays down a framework for future smart water implementation.

In Arizona, Global Water Resources, a regulated water utility serving 120 000 people, has responded to this legislation by providing customers with its proprietary Fathom smart water meter, which is designed to provide comprehensive real time usage data. The Fathom system is also being sold to other utilities.[14]

In Arizona a survey of water conservation measures in 2010 found that 5 of 15 communities offered rebates relating to smart irrigation. These included rebates of USD 22-100 for irrigation audits and rebates of USD 30-250 for domestic lawn smart irrigation systems and in one case USD 5 000 (or one third of total cost) for commercial smart irrigation upgrades. Smart metering was not covered by the survey.[15]

Mandating smart municipal irrigation

Smart irrigation for gardens, parks and recreation has been supported by state level legislation. In Texas, the HB 2299 bill relating to equipment used for irrigation systems came into effect in 2008. The 2007 bill calls for irrigation devices to be fitted with smart controllers and water and was modified to call for automatic shut off systems for periods of rain and frost from 2011. California's AB1881 states that from 2012, all new irrigation devices in the state must have smart controllers.[16]

Data privacy and public health concerns as a constraint

Smart water is hindered when policies limit the collection of data from consumers, for example regarding privacy laws and data security requirements. One contrary trend is the decision by the California PUC to allow customers to opt-out from using smart meters over concerns about electromagnetic fields (EMFs) generated by the meter radio transmitted. This reduces the utility of the "smart" meters and drives up costs as customers have to have their meters manually read.

4. Lessons learned

The information above highlights how a combination of policies can contribute to the diffusion of innovation in water management. It also signals how innovation can diffuse as an unintended consequence of initiatives in other areas. The main policy lessons are as follows:

- Initiatives on smart water systems have been more effective when articulated as part of a strategic approach to water management. Malta and Jersey have adopted national smart metering strategies to address water shortages while avoiding the costs of desalination. However, such national strategies remain scarce, as smart water generally has a low profile in water policy development. Efforts have often been piecemeal and disjointed, and opportunities for synergies have been lost (for instance for co-ordinated roll-out of water and power meters).
- Smart water systems combine technical and non-technological innovations. ICT can only make water systems smart when the information collected can be used to change the way water is consumed and managed. This requires that technology developments be combined with innovations in the business models of water utilities and water use at the end of the pipe.
- Technical developments in smart water systems do not necessarily originate in water policies. This is particularly the case in Korea, where smart water innovation has been driven by ICT developments, and in Malta, where one of chief drivers was EU legislation on smart electricity metering and the realisation that smart water metering costs would be reduced by installing both systems simultaneously.
- Supporting the development and deployment of smart water systems requires the co-ordination of an array of policies. Accompanying measures can play an important role: water pricing (in Jersey; or Southern Water's transition scheme in England), education and information (in Australia), and competition for non-domestic consumers (in Scotland).
- Public-private partnerships, although not very common in the sector, have proved to be effective. The Netherlands, Singapore, Israel and Ontario have worked on linkages between their utilities, industries and universities to develop innovative water enterprises both for the home and international markets. The response from the private sector comes from several perspectives. Companies active in the sector such as Veolia Environnement (m2o mobile data joint venture with Orange and the

Veolia Innovation Accelerator), Suez Environnement (Blue Orange) develop and deploy innovations. Selected utilities (e.g. American Water Works, Global Water Resources) adopt smart water grids and metering early, as part of their corporate marketing strategy, along with other utilities using smart water management and monitoring to improve services and reduce costs. New comers in the industry such as General Electric (Eco Imagination) and IBM (Smarter Planet) have developed a presence in smart water technologies based on their traditional competences and their ability to develop innovation in alliance with their clients. These companies are international in their scope and are seeking to have their offerings broadly adopted. Large companies often partner with smaller ones (e.g. from the ICT sector), to combine innovation with their market power.

- There are opportunities for international co-operation. Common standards for hardware and software are needed to encourage the international deployment of innovative technologies. This will become of increasing importance as various aspects of smart water management, which are currently operated independently, are integrated. The International Telecommunications Union reviewed smart water communications standards in 2010 and noted that standards for smart meter communications, smart meter reading devices and smart grid customer interfaces only occurred at the national level at the time.[17] Smart metering standards are under consideration at the European level, including the EN 13757 standard for data transmission between a smart meter and a data concentrator. International standards for water data-communications and geographic information systems are being developed by the ISO and the WMO.

Two sets of policy issues on the development and deployment of smart networks in general are relevant for smart water systems. They open new avenues for policy and corporate strategy, as they relate to market power, and the distribution of value created by big data management.

One relates to the way data is collected and managed from networks and/or from consumers (OECD, 2012). Several options might materialise depending on the distribution of market power and influence in smart networks and services. New global players might emerge that manage smart systems for millions of people and businesses. In a context where smart applications are plentiful and originate in a variety of sectors (energy, water, transport, health and safety, media, etc.), a smart home can be managed as an integrated whole. This creates business opportunities for telecommunication operators, IT companies, or metering companies: in developing cloud services for smart metering, they all are potential competitors in this market. Alternatively, smart systems can come with their own management interface, which may be controlled by a different company. There is ample evidence of energy and telecommunication companies moving up the value chain, not just delivering connectivity and energy, but also actively managing devices in the home or in businesses.

Another set of issues relates to the value of data generated by smart networks and competition for access to that data. As smart systems are deployed across economies and societies, the policies surrounding access to data they generate will be of great influence on the value of the systems to these economies and societies. In some cases, the systems will be controlled by public sector entities. In others they will be controlled by private sector entities that have to conform to public requirements. Public-private partnerships are yet another option. The value of such data and the benefit to society will depend on who has access to these data, and how data can be transferred to third parties. This can also drive additional innovation in services based on information.

5. Current limitations of policy support for Smart Water Management in Korea

SWM in Korea is a remarkable development, which creates value for water management authorities and Korean society. It has ample potential in Korea and abroad. It creates business opportunities for K-water and business partners. So far, the deployment and diffusion of SWM in Korea has relied heavily on investment and support from K-water. Further monitoring is needed to assess how the potential benefits can generate incentives for further diffusion, in Korea and abroad, by creating a demand that can be covered by stable revenue streams.

5.1. Dam management

So far, dam operation and management systems only cover infrastructure operated by K-water. In June 2016, the Korean government decided to unify the operations and management of hydropower dams. As a consequence, hydropower dams and multipurpose dams are managed conjunctively (see Box 4.2).

Box 4.2. **Unification of dam management, as a response to climate change**

In June 2016, the Korean government decided to unify the operations and management of hydropower dams for efficient water management: the hydropower dams (Paldang, Cheongpyeong, Hwacheon, etc.) managed by Korea Hydro and Nuclear Power (KHNP) are going to be operated in conjunction with K-water's multipurpose dams in terms of discharge quantity and dam water level.

Hydropower dams are owned and operated by KHNP to maximise hydropower generation. Multipurpose dams are owned by the government (MoLIT) and operated by K-water; they discharge water year round to serve several purposes, such as flood prevention, water supply and power generation.

A joint operation is expected to generate multiple benefits. First, additional flood control capacity can be secured and the flood discharge can be reduced through real-time dam operations. Second, water supply can be secured without building dams by reserving unused water in the dams upstream. The additional secured volume is estimated at 540-880 million m^3/year, equivalent to 5 times Yeongju dam, which costed KRW 1.1 trillion. Third, it can contribute to improved water quality downstream by flushing discharge of upstream dams.

To effectively operate and manage dams in the future, the gradual diffusion of smart water management to all sectors is recommended. This would allow all sectors to benefit from improved water management and facilitate joint management of water resources, in particular during times of scarcity. Partial coverage limits the potential benefits of SWM as a tool to mitigate flood and drought risks; therefore, it would be interesting to liaise with MAFRA and KRC during the development stage to explore opportunities for collaboration.

Some of the questions to be explored jointly include: How can SWM be rolled out to help operate the network of small dams for agriculture? What are the potential gains for the farming community? Can these benefits provide a new stream of revenue (public or private) to finance the costs of investing in and operating SWM for the management of flood and drought risks in the agriculture sector? What are the potential benefits of such extension for the broader community (e.g. when agriculture water can supply cities in cases of severe droughts, or agriculture land can provide a buffer to manage flood risks)?

SWM relies on extensive coverage by ICT and the capacity to maintain and operate the system. This may be appropriate in the Korean context, but may create difficulties in developing countries. The diffusion of SWM abroad should be combined with the construction, operation and maintenance of ICT facilities considering the unique economic and social situations of each country. It also requires investment in water infrastructures to move water around as needed. Experience with the diffusion of green technologies suggests that diffusion will be greater if target countries have the capacity to adapt and adjust innovation to local conditions: not simply importing the technology, but combining it with local knowledge and know-how.

5.2. Management of water demand and water services

SWM in Korea is a K-water-driven, engineered response to two problems. One is how to minimise non-revenue water, especially in municipalities. Chapters 1-4 noted the constraints faced by supply augmentation in Korea at subnational level. Minimising leaks in Korean cities can help make a better use of available resources, while saving costs of investment in new storage capacities. This is an issue for Korean cities, which strive to keep water bills low and face investment needs to serve a growing urban population and renew existing water supply infrastructure. The other problem is to increase revenues from water supply services, in a context where households tend to drink bottled water as a sign of distrust vis-à-vis water suppliers.

SWM solutions should not exclude users' priorities and concerns. In OECD countries, users usually are not concerned about non-revenue water, especially when water tariffs are low. Some level of competition may be beneficial for Korean society, when alternative developments can stimulate creative responses to consumers' needs and avoid lock in in one particular technical trajectory. In that context, the forthcoming initiative by the Korean government to accelerate the renewal of water supply and sanitation infrastructure can contribute to the further deployment of K-SWM at local level.

There is little incentive for water users to minimise non-revenue water, in particular as tariffs for water supply are very low and well below affordability thresholds. Efforts to minimise non-revenue water will resonate with users' priorities and concerns when water tariffs better reflect the full cost of supplying water, including the cost of investing in and operating and maintaining the water supply infrastructure, the environmental and opportunity costs of abstracting water and preventing other uses (for industry, the environment, or else). Similarly, a municipality or utility will only be able to finance K-SWM when water savings compensate investment and operating costs, that is, when revenues from water tariffs tend to reflect full water supply costs. This explains why the diffusion of SWM in Korea is directly related to issues discussed in Chapter 3, and to other issues related to tariffs for water supply and sanitation services (an issue not covered in this report).

5.3. Governance

K-SWM is a large-scale technology-intensive strategy with the ambition to be an "integrated water management model that ensures the stability, safety and efficiency of water […]" and to provide smart water services that rely on two-way communication to "improve the response time to meet consumers' needs" and that would promote "consumer-oriented water management" (K-water, 2016). The realisation of these objectives will not only depend on technological performance, but equally on whether the appropriate governance conditions are in place to facilitate SWM uptake, in terms of institutional co-ordination, capacity building and stakeholder engagement. A careful look at current efforts to implement K-SWM however sheds light on a number of governance drawbacks that, if not addressed, are likely to stymie the realisation of SWM objectives and expected benefits.

Sectoral integration

SWM in Korea aims at integration of different steps in the water management cycle (from dam management and flood control, to water supply and conservation). It is expected that ICT will help reconcile existing water management practices that are conducted separately with the hope of improving the efficiency and effectiveness of water management.

As SWM in Korea develops, it can achieve this ambitious goal when:

- Its coverage extends and includes agriculture dams, in co-ordination with similar initiatives from Korea Rural Community Corporation.
- Pilots fully reflect the integrated approach. It should be noted that from a business development perspective, K-water should develop a suite of tools that contribute to SWM and can be operated independently. To some extent, this has already been done, e.g. different components (e.g. data collection, modelling, communication, operation and maintenance) are promoted or implemented separately; and there is a lot of value in combining these components responsively to address specific issues or priorities raised by potential customers (e.g. cities) in Korea and abroad. This may support the deployment of SWM in a stepped approach, which may be more palatable for customers than a systemic one.
- Stakeholders (essentially households) are involved in the definition of services that respond to their demand and needs.

Capacity gap at the local level

Implementation of SWM in Korea would benefit from capacities of local governments to understand and support the objectives of K-SWM, actively promote it and contribute to its implementation. At this stage, it is not clear whether such capacities are widely spread across Korean municipalities. The pilot cases are an opportunity to systematically assess the scientific, technical and organisational expertise needed to shoulder implementation of SWM in Korea. Pilot cases carried out up to now have focused on cities that have signed a consignment contract with K-water for the management of their water services. In such contexts, K-water – not local government staff – was able to practically deploy SWM.

The further deployment of SWM would benefit from an assessment of the capacities required by local governments that have not signed a consignment agreement with K-water. K-water would benefit from exploring demonstration cases with cities without a consignment contract, in order to gauge the expertise required of local water staff. It would also provide a "reality check" on the feasibility of deploying SWM by identifying critical capacity gaps, and develop the specific training and knowledge needed to bridge these gaps.

Customer involvement

SWM has the potential to combine technological advancements with social innovations, but it will imply that water users be more actively involved in the definition of the services they will benefit from. While SWM has the ambition to achieve "consumer-oriented water management" (K-water, 2016), at this early stage, water users are limited to a passive role of "technology takers" when it comes to its design and implementation. So far, consumers' interest in water supply and water service tend to be limited and low, and K-water has had to figure out what kind of information might be useful from a final user's perspective. Sustained efforts are required to better understand the kind of information, service or benefit most valued by final users.

This would entail a two-way relationship between K-water and water users. Thus far, the communication has primarily been towards water users, in a typical "technology push" approach (see OECD, 2011), which takes the form of one-way information-sharing, for example via smartphone applications and information boards in building hallways to provide data on water quality and leakage rates. Such efforts are welcome: they can make customers more knowledgeable and sensitive to specific water issues, and can trigger engagement.

Customer involvement would benefit from other forms of engagement, ranging from communication (i.e. information sharing) to consultation, participation, representation, partnerships, to co-decision and co-production (see OECD, 2015, for more information). As a first step forward, K-water could provide final users with an opportunity to signal what kind of information would be useful for them, under which template, and the potential benefits they expect from improved knowledge and data about water availability, quality and use.

SWM has the potential to go a step further and explore how water services, both in their design and provision, can be leveraged through new behavioural insights from water users. Consumer perspectives may not automatically coincide with the perspective of K-water or local providers. As such, the inclusion of consumer perspectives in the early stages of the innovation process would offer K-water and local authorities valuable information about users' needs and expectations.

The Smart App developed by the Smart Water Grid Research Group and K-water provides a good illustration. The SWG Research Group is piloting this app in Yeongjong-do, Incheon city with a population of about 20 000. The application provides remote meter reading service, water usage options, two-way information sharing of water use, real-time outage notification, and customer support. K-water's App provides real-time water quality information for each supply process, CCTV images of apartment water tank, water usage and billing information, and service and complaint application functions. It is being used in two K-SWM pilot cities: Paju and Goryeong. The question remains: do water users need this information? What best use can they make out of it? How does this information enhance users' experience with the water service and its provider?

For stakeholder engagement to strengthen legitimacy, acceptability, and accountability of the water service provider, it is essential to give stakeholders a voice and to accommodate their inputs effectively and regularly. In addition, the degree, scope, quality, and efficacy of the incorporation of civic participation must be seriously considered when evaluating the performance of governmental institutions and public officials. It will require that the technology developments of SWM be combined with innovations in the business model of local authorities, one that supports a new, pro-active relationship with water users beyond information sharing. In addition to increasing technical performance, smart meters, ICT and tools assessing customers' behaviour under various scenarios can be used to improve customer control and choices and to change the way water is consumed. Ultimately, it will lead to better resource efficiency, planning and management of water services. By aligning technologic and consumer-driven ambitions, SWM can allow water users to have a say in defining the future of the service and better controlling the way they use water.

6. Options for reform

In Korea:

- Promote an integrated approach to water management in Korea. This may be a role for the Water Management Committee under the Prime Minister.
- Engage in a discussion with MAFRA, KRC and farming communities at both national and basin levels on the potential benefits of extending SWM to agriculture reservoir use, while paying due attention to the distinctive features of agricultural water management. Explore how the benefits can be turned into revenue streams that finance the investment and O&M costs of such an extension.
- Work creatively with municipalities, industrial and domestic water users to develop value-adding services built on the new data available (e.g. tariffs for water supply that reflect peak hours). Some competition for the development of services that add value to final water users may be considered.
- Assess capabilities and know-how required for local authorities to demand and manage SWM, such as during the pilot cases, and contribute to building such capacities, e.g. through specific training.
- Explore where modelling capacities can minimise needs for new data.
- Continue to make the case for water pricing instruments that reflect the value of water (scarcity and opportunity costs) and for water tariffs that reflect the full costs of supplying water and sanitation services. This is a requisite to create a demand for SWM in Korea.

K-water has a leading role to play for several reasons. First, it masters the technology and its future developments. Second, while it reports to MoLIT, it liaises with all relevant institutions active in water resources management and water supply (ME, MAFRA, cities). Moreover, it has a business model to roll out the technology and associated services, both upstream (working with MAFRA on irrigation dams) and downstream. To tap its full potential, K-water would benefit from enhancing its experience with final water users (typically households) and its capacity to reflect needs and expectations expressed by them. It would also benefit from diversifying its business model, by designing a range of services to be provided to municipalities which do not have a consignment contract.

Abroad, K-water will be better placed to export SWM if target countries are able to adopt and adapt the technology and related functions to their local conditions and know-how. Therefore, K-water, possibly in partnership, would benefit from:

- Building capacities in target countries to adopt *and adapt* SWM. This requires that K-water abandons some sort of control over its invention. KOICA and multilateral agencies can play a role, as they work directly with national authorities abroad.
- Developing standards for ICT and water technologies (with the Ministry of Trade), to contribute to interoperability of several systems domestically and to grow global demand internationally.
- Rolling out SWM in combination with related technologies (e.g. smart metering for energy consumption).

Notes

1. Collecting more systematic information about policies to support *water-related* innovation is beyond the ambition of this report.
2. Renner S. et al. (2011) European Smart Metering Landscape report, *Intelligent Energy Europe*, Vienna, Austria; Goldstein H. (2010) Malta takes control, *Spectrum*, IEEE, 47 (6), 65-71, June 2010; Revolt A.E. (2010), The Trade-Off Between Water and Energy: Malta's Multi-Utility Smart Grid Solution, *Worldwatch Institute blog*, 8 October 2010.
3. MRRA (2009) *National Strategy for Policy and Abatement Measures Relating to the Reduction of Greenhouse Gas Emissions*, Ministry for Resources and Rural Affairs, Government of Malta.
4. Cleantech Group (2011) Ontario Global Water Leadership Summit. Cleantech Group, San Francisco, USA
5. André Struker and Marc Havekes, Waternet, Personal communication, 2012; Robert Riley, Senior Principal, Accenture, Personal communication, 2012
6. Cleantech Group (2011), *Ontario Global Water Leadership Summit.* Cleantech Group, San Francisco, USA
7. Lloyd Owen D A (2012) *Singapore – a holistic approach to sustainability.* Paper presented to the Singapore International water Week, July 2011
8. Seth Cutler, Research Analyst, Environment (Water) Markets, Frost and Sullivan, Personal communication, 2012.
9. Ken MacDonald, Ofwat, *Presentation at the SBWWI 2010 Metering and Leakage Seminar –* New Challenges: New Solutions.
10. Ofwat (2011) *Exploring the costs and benefits of faster, more systematic water metering in England and Wales.* London.
11. John Crooke, Southern Water. *Presentation at the SBWWI 2010 Metering and Leakage Seminar* – New Challenges: New Solutions.
12. Chad Staddon, Department of Geography and Environmental Management, University of the West of England, Bristol, Personal communication, 2012; WICS (2010), *Competition in the Scottish water industry 2009-10*, Water industry commission for Scotland.
13. Mark W. LeChevallier, Director, Innovation and Environmental Stewardship American Water, personal communication, 2012.
14. Graham Symmonds CTO and SVP-Regulatory Affairs and Compliance, Global Water Resources, personal communication, 2012; *Are you leaking water or data?* (www.gwresources.com/news/Documents/publications/watercanada-are_you_leaking_water.pdf); *Using the Smart Grid for Water to Convert Non-Revenue Water to Found Revenue Water: Controlling Leaks of Data, Drops and Dollars.*
15. Western Resources Advocates (2010) *Arizona Water Meter: A Comparison of Water Conservation Programs in 15 Arizona Communities*, WRA, Boulder, Co, USA.
16. Updated Model Water Efficient Landscape Ordinance AB 1881; www.water.ca.gov/wateruseefficiency/landscapeordinance/.
17. ITU (2010), *ICT as an Enabler for Smart Water Management,* ITU-T Technology Watch Report, ITU Geneva, Switzerland.

References

Analytica Advisors (2017), *2017 Canadian Clean Technology Industry Report*, Ontario, www.analytica-advisors.com/publications.

Daniell, K.A., Coombes, P.J., White, I. (2014), Politics of innovation in multi-level water governance systems, *Journal of Hydrology*, 519 (2014) 2415–2435, Elsevier, doi:10.1016/j.jhydrol.2014.08.058.

Environment and Climate Change Canada (ECCC) (2016b), *Achieving a Sustainable Future: A Federal Sustainable Development Strategy for Canada,* Ottawa, http://fsds-sfdd.ca/index.html#/en/detail/all/.

European Innovation Partnership on Water (2014), *Barriers and bottlenecks for innovation in the water sector*, Final Report, European Commission.

Haley, Brendan (2016), *Getting the Institutions Right: Designing the Public Sector to Promote Clean Innovation,* Smart Prosperity Institute, http://institute.smartprosperity.ca/library/publications/getting-institutions-right-designing-public-sector-promote-clean-innovation.

Invest Ontario (2016), *Clean water technology in Ontario: A global hub of research and innovation*, Toronto.

K-water (2016), *New Era of Water Resources Management with the Smart Water Management Initiative*, K-water, Daejeon.

K-water (2015), *Water for a Happier World, K-water Sustainability Report 2015*, K-water.

OECD (2016), *G20 Innovation Report 2016*, Report prepared for the G20 Science, Technology and Innovation Ministers Meeting, Beijing, China, 4 November 2016, OECD. www.oecd.org/innovation/G20-innovation-report-2016.pdf.

OECD (2015), *Water Cities – Ensuring Sustainable Futures*, OECD Studies on Water, OECD Publishing, Paris. http://dx.doi.org/10.1787/9789264230149-en.

OECD (2012), *Policies to Support Smart Water Systems: Lessons from Country Experience*, unpublished paper by David Lloyd Owen commissioned by the OECD.

OECD (2011), *Better Policies to support Eco-innovation*, OECD Publishing, Paris. http://dx.doi.org/10.1787/9789264096684-en.

OECD (2005), *Governance of Innovation systems, Vol.1: Synthesis Report*, OECD Publishing, Paris. http://dx.doi.org/10.1787/9789264011038-en.

Chapter 5

Water allocation

This chapter reviews the situation with water allocation regimes in Korea. It highlights the current limitations and the consequences for water management in the country, and it points at options for reform. The level of ambition of this chapter is limited, as the issue cuts across administrative and sector boundaries and requires extensive consultation with water users and water policy makers from a range of perspectives. Such a review is beyond the objective of this project, but would benefit from a dedicated water policy dialogue in Korea.

1. Past heritages and current status of water entitlements

Up until the mid-1950s, Korea essentially was a rural community involved in farming, predominantly rice cropping. This history has a strong legacy in current water management institutions, practices and policies. It explains that agriculture has developed its own system of small dams and reservoirs, to secure access to water, independently from other users. However, farmers, as beneficiaries of customary water rights, are exempted from regulations and economic instruments that drive water use in Korea, typically water entitlements, and the river water use fee.

In principle, since the River Act in 1961, the right to use river water in Korea is granted by the government, via Flood Control Offices, and all water uses should be registered and permitted. Flood Control Offices review applications for new or additional water use. They inquire whether new claims affect vested amounts. If the new water use does not affect previous entitlements, claimants will be granted new rights and be charged the river water use fee. In other cases, the claimant will be invited to procure water from a dam upstream (through an agreement with K-water), and the dam water tariff will apply. As noted in Chapter 3, the river water use fee and the dam water tariff are the same, but revenues are collected by different entities (local authorities for the former and K-water for the latter).

Water entitlements that existed prior to the River Act in 1961 are considered customary rights. These rights usually are small amounts of water used for agriculture and are acknowledged by the Civil Law. It remains to be seen how much the aggregate amount affects water availability for other purposes in river basins. Customary water rights have been analysed in the last decade, and data is currently being registered. Their impact on other uses has not yet been reviewed by Flood Control offices.

For small rivers, a dedicated law sets out conditions of access to water, under the responsibility of local governments. They are regulated under the Small River Maintenance Act, while national and local rivers are regulated under the River Act.

All registered water uses can formally be reviewed by River Adjustment Councils, which report to Flood Control Offices. At the moment, the four rivers are closed for new allocation, meaning that no river water is available for additional uses. As a consequence, new users need to apply for dam water. In the current context, dam water is available in every basin, and no revision of existing entitlements for addressing additional water demand has been required.

A pending issue relates to the un-used water entitlements. Water right holders use 35.2% of their entitlements on average. The ratio climbs to 52-53% for non-agriculture water. This situation reflects a tendency to allocate water rights above and beyond actual needs. This can be seen as a buffer for times of scarcity. This is also an inefficient way to allocate water, as water entitlement holders have little incentive to use water wisely, while new comers may experience difficulties to access water they could use for valuable activities. Aligning permits with actual water uses would make some river water available for new uses, potentially creating more value with existing water resources.

Minimal flows for river maintenance are determined at different critical points of rivers according to Article 51 of the River Act. However, these flows a) may not guarantee that environmental services will be maintained, if they do not fulfil ecosystem requirements; and b) may not be the best use of available water: more water may be retained for environmental purposes than is required to achieve the desired outcomes, thus unnecessarily limiting the size of the consumptive pool. OECD countries apply more sophisticated methods to determine environmental and minimal flows and the best use of available water, which may be beneficial for Korea (see below).

Setting environmental flows – Definitions and methodological options

The importance of environmental flows is now widely recognised and legislation in many countries enshrines the requirement that environmental water needs be considered as part of the allocation process. This section draws on OECD (2015a, 2015b).

The definition of e-flows – results from the OECD survey

A significant majority (76%) of respondents to the OECD survey indicated that minimum environmental flows are defined. A wide range of methodologies to do so was reported. For example, in Israel, in some places a minimum quota of water has been set aside and must be allocated to ecosystems. In Slovenia, the ecologically acceptable flow is set depending on the type of water use and type of ecological needs. In England and Wales, environmental flow indicators are used as an indicator of the flows required by the environment. In Portugal, minimum environmental flows are determined on a case by case basis. In China, the warning-level river flow against the drying out of a downstream river course shall not fall below 200 cm^3/sec at Xiaheyan hydrological stations. In the Murray-Darling Basin, Australia, the Basin Plan limits water use at environmentally sustainable levels by determining long-term sustainable diversion limits for both surface and groundwater resource. A key component of the Basin Plan is the environmental watering plan, which coordinates all environmental watering across the Basin.

Of the examples indicating that minimum environmental flows/sustainable diversion limits are taken into account, 82% take freshwater biodiversity into account in the definition of e-flows and 64% take terrestrial biodiversity into account. For example, in France, the minimum biological flow and the reserve flow required are based on the observation of ecological needs.

Options to set environmental flows

Basic hydrological methods to determine e-flows, like the Tennant method, rely on the establishment of relationships between flow and ecology. In the case of the Tennant method, the original relationships were based on observations of how stream width, depth and velocity varied with discharge on eleven small mountain streams in the United States. Because of the wide natural variation in river hydrology and ecology throughout the world these relationships are unlikely to have universal application.

Environmental flow assessments are ultimately an input to a socio-political process. While the natural sciences can provide information on what the implications will be for different parts of the environment of changing the flow regime, it is a socio-political decision as to what ecosystem services should be protected, and hence what environmental flows should be provided.

There are various methods available for determining environmental flows, including hydrology-based methods, which can provide useful information to decision-makers. Ultimately, the nature of the assessment process should be based on the complexity of the system being considered and the risks associated with changes to the flow regime.

Environmental flows should be set, at the basin scale, to achieve clearly defined environmental outcomes. Flow requirements should be determined based on an understanding of the flow regime, its role in contributing to river health and environmental services, and the risks associated with changes to the flow regime. Importantly, research shows that different elements of the flow regime serve distinct purposes when it comes to maintaining ecosystem processes and services. Environmental flows should thus consider the timing, frequency, duration, and magnitude of flows that are required to achieve the desired outcomes.

Reserving a gross volume of water for environmental purposes when setting reference flows is likely to support the maintenance of base flows. However, other measures may be required to ensure the required environmental flows are achieved. For example, reservoir release rules may be necessary to achieve medium or high-flow objectives at the required times, for example to ensure the higher flows required for the movement of sediment, to periodically inundate wetlands, or to trigger fish spawning or migration.

2. Limitations

The coexistence of water entitlements acquired before the construction of dams and reservoirs (customary water use and vested amounts of water) and of more recent ones (right to use dam water, or Permission to Use River Water Rights, which depend on an assessment of water availability) is dysfunctional.

It creates legal disputes – which are litigated in courts – and inequities as regards who can access water and under which conditions. Typically, senior right holders benefit from a rent situation and only use a portion of their entitlement, while new comers, who may be able to use water wisely, may be denied access to river water. The prevailing response has been supply augmentation via additional dams and reservoirs. Aligning water permits with actual volume of water use can contribute to creating more value with existing water, without additional construction.

The system is also economically inefficient – as it fails to allocate water where it creates most value and is socially inequitable – as risks of water scarcity are inequitably shared among users. The water allocation regime does not reflect the differentiated capacity of water users to address risks of scarcity.

The allocation of water for river management creates rigidities and potential inefficiencies, as mentioned above: the River Water Adjustment Council can adjust the amount of water allocated to different uses, but transaction costs can be high and few cases have been brought up so far.

The system adjusts to variations in water availability, but only at high transaction costs: it takes a lot of time to negotiate tailored arrangements, which are politically and socially sensitive, and potentially economically damaging (when farmers, businesses or investors lack certainty about future access to water). In cases of scarcity, the River Act provides for a redefinition of priorities: i) domestic water; ii) industrial water; iii) agriculture water; iv) other purposes. As a consequence,

- Environmental flows can be suspended.
- Some water uses in a lower priority can be banned, creating more inequality to the management of water-related risks.
- K-water can occasionally buy water from the agriculture sector to secure supply water for domestic and industrial use. KRC (which administers agriculture water) has a provision to generate revenues from selling water to non-farmers. However, farmers can resist this diversion of agriculture water for other uses.

International experience indicates how welfare gains can derive from water allocation reforms without (or with only minimal) additional investment in water infrastructure. For instance, optimising water allocation in times of scarcity can minimise the negative impacts of droughts on economic production, and lessen the need for investment in supply augmentation.

Governance responsibilities related to water allocation are split across central and sub-national institutions, and differentiated whether they concern river water, dam water, multi-regional water or groundwater. ME, MOLIT (via K-water) and local authorities are responsible for domestic water supply; local authorities and MOLIT (via K-water) are in charge of industrial water supply; while MAFRA and KRC deal with the agriculture use. Flood Control Offices issue permits for the use of river water, while local authorities oversee the allocation of groundwater. In addition, allocation for these four water sources is prescribed by different legal frameworks, i.e. the River Act for river water, the Groundwater Act for groundwater resources, while the Basic Plans of Dam Construction and the Basic Plans for Waterworks Installation and Management dictate the allocation of dam water and multi-regional water respectively.

Second, Korea's water allocation regime would benefit from properly factoring equity goals. At the moment; not everyone has fair opportunities to access water resources: Article 49 of the River Act stipulates that water shall be used "within the limits of not impairing the individual right of others and public interest, while not hindering water management, and shall be distributed in a way to evenly benefit all people of the nation". However, in practice, water right holders use only 35.2% of their permitted water abstraction on average. On top of this, water entitlements that are not used are lost (i.e. "use it or lose it" approach). Together, these two factors create an injustice between those who historically have been allocated entitlements, and new entrants looking to access water resources. As such, the current allocation regime falls short from promoting the interests of new comers. The situation reflects a tendency to allocate water rights above and beyond actual needs, which in turn is inefficient at allocating water for activities that create the most value for the community. Also, the potential consequences of water scarcity are unevenly distributed: senior water right holders have little incentive to use water wisely and make more water available to others, while junior right holders can be banned in cases of scarcity.

Moreover, not everyone is granted the opportunity to participate in water allocation decisions: while each KRC local office has a consultative council of 12 representatives from the agricultural community that take collective decisions. For what concerns water allocation for domestic and industrial purposes, this should be the role of the multi-stakeholder River Water Adjustment Councils, according to the River Act. The River Water Adjustment Councils were organised in 2015 for the Han and Geum Rivers, and in 2016 for the Nakdong and Yeongsan Rivers. To date, the Han River Basin Adjustment Council has met only twice; the other councils have met only once. Water is allocated as a public resource by MoLIT in consultation with related central administrative agencies and local authorities. Attention should be paid to the value added of user-based allocation, and its potential flexibility to adapt water allocation patterns to local needs and to the distinctive capabilities of various water users to adjust to water scarcity. Because those directly involved in water use, whether for home consumption or industry, often have more information than central governments on local conditions and the capacities of different users to manage water scarcity risk, they should be offered the opportunity to contribute to water allocation in practice. Water allocation regimes should build on such capabilities. Developing a communication and engagement strategy to inform and involve water users in the development and implementation of water allocation regimes can help to build a sense of the value of water entitlements and the importance of a strong, yet flexible, allocation system to protect the long-term interests of water users.

3. Options for reform

A recent review of water allocation regimes in OECD countries and beyond indicates that since the risk of shortage is dynamic in both the short-run and the long-run a well-designed allocation should have two key characteristics: i) it should be robust by performing well under both typical and extreme conditions, and ii) demonstrate adaptive efficiency with the capacity to adjust to changing conditions at least cost over time.

Allocation regimes need to be tailored to specific conditions. In the early stages of developing a water resource, or where the risk of shortage is low, a relatively simple design can be used with decisions made conservatively to avoid over-allocation and over-use. As scarcity increases and the value of water use rises, the benefits of a more elaborate allocation regime increases.

To align with international good practices and make the best use of available water resources, in normal times and in times of scarcity, Korea could:

- Embark on a full OECD review, using the *OECD Water Allocation Health Check* at national and basin level (see Box 5.1), and building on recent experience of a similar review in Brazil. The review will figure out how water allocation regimes can be streamlined in Korea so that they address equity issues (unequal allocation of risks of scarcity), environmental issues (insuring that ecosystems and the services they provide are conserved), and economic ones (allocating water where it creates more value), while avoiding building additional infrastructure to augment supply. It will fit well into the paradigm shift encouraged by the Smart Water Management Initiative.
- Strengthen partnership between MoLIT/K-water, MAFRA/KRC, ME/KECO and local governments. They manage distinct but equally relevant and possibly interconnected water supply systems, and water policy instruments. A combined approach of an aggregate water pool could reap substantial benefits in terms of flexibility and economic gains. Looping in the ME would only strengthen the process, as water quantity and quality are linked. There are benefits in looping in MoSF as well as water allocation regimes can drive investment in water infrastructure and depend on economic (in particular pricing) instruments for water management and water supply and sanitation services.
- Consider initiating reform as soon as possible. International experience (see Annex 5.A1) suggests that a) such reforms take time (some 20 years in Australia, or in the UK); and b) they are less costly, economically and politically, when they are initiated while there is still some room of manoeuvre to negotiate with water users and possibly compensate them, than when catchments are already over-allocated and water over-used, and when competition across water users intensifies.

Box 5.1. **The OECD Health check for water allocation regimes**

The OECD has set out a *health check* for improving the performance of existing allocation regimes. The health check is designed as a tool to review current allocation arrangements in a given area. It uses a series of "checks" to identify whether key elements of a well-designed allocation regime are in place and how their performance could be improved.

Check 1. Are there accountability mechanisms in place for the management of water allocation that are effective at a catchment or basin scale?

Check 2. Is there a clear legal status in place for all water resources (surface and ground water and alternative sources of supply)?

Check 3. Is the availability of water resources (surface water, groundwater and alternative sources of supply) identified and possible scarcity well-understood?

Check 4. Is there an abstraction limit ("cap") that reflects *in situ* requirements and sustainable use?

Check 5. Is there an effective approach to enable efficient and fair management of the risk of shortage that ensures water for essential uses?

Check 6. Are adequate arrangements in place for dealing with exceptional circumstances (such as drought or severe pollution events)?

Check 7. Is there a process for dealing with new entrants and for increasing or varying existing entitlements?

Check 8. Are there effective mechanisms for monitoring and enforcement, with clear and legally robust sanctions?

Check 9. Are water infrastructures in place to store, treat and deliver water in order to allow for the allocation regime to function effectively?

Check 10. Is there policy coherence across sectors that affect water resources allocation?

Check 11. Is there a clear legal definition of water entitlements?

Check 12. Are appropriate abstraction charges in place for all users that reflect the impact of the abstraction on resource availability for other users and the environment?

Check 13. Are obligations related to return flows and discharges properly specified and enforced?

Check 14. Does the system allow water users to reallocate water among themselves to improve the allocative efficiency of the regime?

Source: OECD (2015a), *Water Resources Allocation. Sharing Risks and Opportunities*, http://dx.doi.org/10.1787/9789264229631-en.

Annex 5.A1

Selected experience with reforms of water allocation regimes

The OECD has reviewed experience of a range of countries with the reform of allocation regimes (see OECD, 2012). Two cases mentioned below highlight the difficulties related to such reforms.

Making the case for water abstraction reform in England and Wales

The example of England and Wales demonstrates the value of making a clear case for change, including ensuring that the shortcomings of the current system are widely understood. To support the proposals for abstraction reform set out in the White Paper "Water for Life" (2011), the United Kingdom Environment Agency developed "The Case for Change", which assessed the potential impacts on water availability of climate change, couple with increased demand for water from population growth and energy supply needs. The assessment considered 11 emission scenarios from the United Kingdom Climate Projections (UKCP09) and their potential impact on river flows.

The conclusions were varied, but suggested that average summer river flows could reduce by up to 80%. This reduction would not be offset by increased winter precipitation. Attempting to maintain current (aspirational) levels of environmental protection would reduce significantly the water available for abstraction. Similarly, demand pressures under a range of socio-economic scenarios would pose a threat to environmental limits. The analysis considered averages; drought events would have greater impact. The south and east of England is already water stressed from a combination of low rainfall and high demand. But the wetter west and north could not be relied upon to make up any shortfall: the analysis suggested that these catchments would be most affected by climate change consequences.

The "Case for Change" spelt out the future challenges and pressures on water resources. However, in order to understand how the allocation system should be reformed to respond to those challenges it was important to understand its shortcomings. In December 2013 the government published "Making the Most of Every Drop", a consultation on abstraction reform, which set out why reform was necessary:

- The current system does not systematically link access to water to availability. Only a quarter of licences have controls to stop abstraction to protect the environment or other abstractors during periods of low availability. Conversely, the system struggles to allow additional water to be taken during higher flows.
- The system does not help abstractors to trade available water effectively, and to provide price signals to promote efficient water management.
- Abstractors are not currently incentivised to manage water efficiently.

- Much of the water (generally less than half) that is licensed is not actually used. This potentially denies access to others.
- The current process to change most licences that are causing damage to the environment is expensive and time consuming. As the climate changes and flows reduce or become more variable, more licences are likely to require changes, making this problem much worse and more expensive.
- The system fails to incentivise abstractors to manage risks from future pressures on water resources.

The government stated in its consultation that "these weaknesses may constrain economic growth due to reduced resilience and getting sub-optimal economic value from available water, while not efficiently protecting the environment". Officials from the Department for Environment, Food and Rural Affairs (Defra) worked closely with the Environment Agency, and representatives from a wide range of other organisations, in order to fully develop the policy options for reform.

Factors stalling implementation of water allocation reform in South Africa

The Water Allocation Reform programme in South Africa recognised early on that getting the pace of reform right was key: move too slowly and you are likely to see radicalisation of policy as the political imperative for redress increases, move too fast and you may threaten the economic value of existing water use, limit the value of improved management of the resource, and increase the likelihood of legal challenges. However, an overly technical and precautionary approach has been taken, and sixteen years since the adoption of the National Water Act (1998), there are still significant challenges in the implementation.

Compulsory Licensing, which had never been implemented anywhere in the world previously, posed particular issues. The concept of Compulsory Licensing was introduced in the Act as a method for the re-allocation of water, primarily from the white minority to the black majority that had been excluded from access to water under Apartheid. This clause enables the Minister to call for all water users and potential water users within a specified area to apply for new water use licences, and for the Minister then, through a consultative process, to re-allocate the water.

A number of factors have made compulsory licencing difficult to implement. The definition of the reserve for ecological and basic human needs also posed a challenge early on. The Act requires that the ecological and basic human needs reserve be determined prior to the consideration of any licence application. However, there were, initially, no procedures in place for the determination of the ecological reserve. The South African aquatic ecologist community set to work in developing such procedures, facing the challenge of making the transition from a scientific analysis approach to developing assessment tools that would serve the purpose of the Act. The need to determine the ecological reserve for significant water resources in the country prior to the consideration of licence applications significantly delayed the issuing of licences for a number of years. In addition, the translation of the reserve requirements into licence conditions was often difficult. For example, where the reserve determination required a fluctuating flow in the river over different months, where a farmer wanted to construct a simple dam with no mechanisms for releasing such fluctuating flows. In addition, the monitoring of the achievement of the ecological reserve has been weak, and so there is a break in the feedback loop between the issuing of licences and the achievement of the ecological reserve.

The first was that all existing water users were required to register their water use with the Department of Water Affairs (DWA), in order to enable the DWA to have a clear record of who was using how much water and where. However, once a process was introduced to check on the accuracy of this registration and the legality of the water use, it was found that an extremely high proportion of the registered water uses were inaccurate, often irrigation farmers over-registering their water use. This required an intensive process of validating the registration, which is still ongoing. In addition, the failure to put in a requirement that the DWA was informed of any transfer of irrigated land-ownership meant that the registration records were out of date where land had been sold. Since Compulsory Licencing was predicated on having a fairly accurate record of existing water use, this delayed the process.

This rigorous reconciliation process is also intensely legal in nature, which may also underpin the hesitancy the DWA has shown in rolling out the process. Legal challenges could delay the process considerably and the DWA may wish to be very sure of their position before tackling large and difficult compulsory licensing processes.

References

Economic and Social Commission for Western Asia (2003), *Sectoral Water Allocation Policies in Selected ESCWA Member Countries: An Evaluation of the Economic, Social and Drought-Related Impact*, United Nations, New York.

OECD (2015a), *Water Resources Allocation – Sharing Risks and Opportunities*, OECD Publishing, Paris. http://dx.doi.org/10.1787/9789264229631-en.

OECD (2015b), *Water Resources Governance in Brazil*, OECD Publishing, Paris. http://dx.doi.org/10.1787/9789264238121-en.

Roa-García, M.C. (2014), Equity, efficiency and sustainability in water allocation in the Andes: Trade-offs in a full world, *Water Alternatives* 7(2): 298-319.

ORGANISATION FOR ECONOMIC CO-OPERATION AND DEVELOPMENT

The OECD is a unique forum where governments work together to address the economic, social and environmental challenges of globalisation. The OECD is also at the forefront of efforts to understand and to help governments respond to new developments and concerns, such as corporate governance, the information economy and the challenges of an ageing population. The Organisation provides a setting where governments can compare policy experiences, seek answers to common problems, identify good practice and work to co-ordinate domestic and international policies.

The OECD member countries are: Australia, Austria, Belgium, Canada, Chile, the Czech Republic, Denmark, Estonia, Finland, France, Germany, Greece, Hungary, Iceland, Ireland, Israel, Italy, Japan, Korea, Latvia, Luxembourg, Mexico, the Netherlands, New Zealand, Norway, Poland, Portugal, the Slovak Republic, Slovenia, Spain, Sweden, Switzerland, Turkey, the United Kingdom and the United States. The European Union takes part in the work of the OECD.

OECD Publishing disseminates widely the results of the Organisation's statistics gathering and research on economic, social and environmental issues, as well as the conventions, guidelines and standards agreed by its members.

OECD PUBLISHING, 2, rue André-Pascal, 75775 PARIS CEDEX 16
(42 2017 46 1 P) ISBN 978-92-64-28166-0 – 2017